ROBOIDS

By

Arnold J. Inzko

To Gabriella, My Wife

A Sci - Fi Novel

1st Edition — June 2017

CHAPTER 1

It is late Sunday evening, during the first week of February 2016. Lena McCabe returned from a skiing trip in Holiday Valley, NY. Her longtime friend and excellent skier, the Persire, Julius Hersey, was her companion. He dropped off Lena at her apartment and then he drove to his home. Lena unlocked her apartment, located in a DC suburb, walked in and stacked her skies against the hallway wall. Then she went to the refrigerator and poured herself a glass of red wine. She drank half of it to warm up her body. She carried the remainder to the bathroom, placed it near the bathtub and then went to her bedroom. Removing her tight ski apparel was difficult, because she was tired and exhausted from the long return trip.

The water for her bubble bath turned off automatically, operated by a sensor, located at the end of the bathtub. Right after she relaxed in the soothing warm water with the glass of wine in her hand, her cell phone rang. *What now*, she thought. On the display she read, Martin Hall. He is Lena's supervisor. *Shit, that's all I need right now,* went through Lena's mind.

"Hi Marty. This is Lena. What's up?" said Lena, submissively and depressed.

"Sorry that I called you on a Sunday night. By the way, how was your skiing trip?"

"Until now, wonderful, Marty. What's so important that you have to call me tonight?"

"It's important, Lena, very important. I can't talk on the phone, but I must see you tomorrow morning at nine in my office."

"See you in the morning, Marty. Good night." Lena closed the cell phone but just before she did she could still hear the click on the other end. As so many times before, Lena is realizing that somewhere in this world trouble is brewing again. Lena was hoping for a relaxed, Monday morning breakfast with Julius but their plans suddenly terminated into oblivion. She called Julius and told him the bad news. Rescheduling breakfast for another day was impractical, because who knows what kind of assignment FBI supervisor Marty Hall has for her.

Lena leaned back in her bathtub, drank the remainder of her wine and reviewed the last few years of her job as special agent of the FBI. During one of her last top-secret investigations, she met Julius when they worked together on a lengthy, dangerous and complicated case. It dealt with the Persires as well as with the now eliminated Ducazians. Originally, both war-like factions came from the planet Nibiru and they used earth as their source for food. They took possession of cattle, tuna and poultry during their raids. The Ducazians also used humans for their sexual pleasures. They captured humans wherever they could, and one of their favorite places was a well-secured monastery. This unearthly practice was the demise of the Ducaz. The Persires, a completely human, lookalike race, formed a partnership with earthlings, with Julius in charge and together they annihilated the Ducaz, chasing them in the deepest part of the oceans as well as in outer space, all the way to the planet Nibiru.

When the Persires, allied with earthlings and finally defeated the Ducaz, the president of the United States allocated a section of land for the Persires, who previously lives in space for thousands of years. It was near the fringes of the Great Basin Desert, covering parts of Nevada and Utah. That was where the Persires landed their gigantic and advanced space city. They built a city next to it and named it New City. Charles Bergen, the seven-foot tall, white-haired Lord Mayor of New City, an ageless looking individual, thrived in his new environment, breathing the warm earth's desert air. He, as well as the others of his race, maintained a healthy and generally peaceful life style. The planet earth was extremely conducive to them. They were practically disease free and their primary concern was that they might contract deceases from humans. With that in mind, his engineers built isolated homes for their constituents and factories to sustain their mode of living. Eventually, they successfully mingled with earth's humans and shared with them their technological knowledge base. Their biggest concern was that humans are war-like people, basically weak and ill-suited to exist on earth. However, the Persires, Julius included, found ways to ally themselves with a selected group of humans. Charles Bergen also decided to initiate a system, aimed at controlling the warlike humans and eventually reducing their disease level.

First, Bergen built plants for items needed to sustain the Persires lifestyle. Then he built a factory to manufacture lifelike roboids for various services. This was one of his most important accomplishments. He started to build them in 2014 and no one could tell the difference between his roboids and Persires as well as humans.

The lord mayor often questioned whether he made the

right decision building realistic, life like machines that are easily mistaken for humans. However, once the assembly lines were in motion there was no way of stopping them. Twice he expanded the operation, because numerous humans of all occupations purchased these roboids. They used them as servants, as baby sitters, as factory workers and on some occasions as teachers; in short, they influenced humans in almost every occupation, because the Persires programmed them to serve everyone's needs. What humans and the Persires didn't realize that these roboids also influenced themselves and eventually had the capability to alter their programs.

The Lord Mayor assigned his only son, Ramon Bergen, to operate and maintain the part of the factory that specializes rebuilding the roboids. Ramon was known to be unreliable and it was difficult for him to maintain discipline and control of this part of the operation. However, with constant supervision, he learned the principles of manufacturing and in due time, he installed the latest security system as well as safety gadgetry. With these advances, he kept good control of the number of roboids that he produced and who bought them.

CHAPTER 2

Julius and Lena present an impressive couple when they walk together in the halls of the FBI, or when they stroll in the Natural Museum of History, observing prehistoric animals or futuristic contraptions. Lena is 5 feet 11 inches tall and carefully uses just enough makeup to accent her facial features. She has a regular workout schedule, to maintain her muscular physique, definitely an advantage in the service of the FBI. She is attractive but not beautiful; she has a sexy and athletic body and walks like a European model. Julius is six feet 6 inches tall. He has the distinct manly, square features of a Persire, slightly tanned skin, with dark brown, curly hair, neatly trimmed and combed back. In his earlier years, he engaged in controlled combat, trying to avoid confrontation with humans. He continues to sharpen his karate skills and maintains his sharp shooters status, practicing regularly at the FBI shooting range. Both are well educated, with their specialty being criminology, and they enjoy each other's company. They first met during a Ducazian raid on a cattle farm in Ohio. Since Julius understood the war-like Ducazian, he took charge observing them, while they harbored their space ship above a cattle herd. They lifted the cattle one by one to their ship, notwithstanding the moo's and groans of the cattle and the bulls. Julius and Lena advanced within firing range, then both of them attacked the Ducazians, firing their ray guns, spoiling their activity. Both nearly succumbed to these poison-spitting, long-limbed and reptile-like creatures, while they mercilessly loaded their spacecraft with cattle, stealing it from the cattle farmer. Fortunately, the agents wore protective apparel. Feeling beaten on a strange planet, the Ducazian retreated to their craft and seconds later it pierced through a thick cloud layer and disappeared.

During their last confrontation with the Ducazians, the Persire, with the help of earthling nearly annihilated all of them, fighting on earth and in the depths of the ocean. Lena saved a small, passive number of the reptiles to further study them and their demeanor. In order to render them harmless, she ordered a surgeon to remove the two poison generating glands, behind their jaws. Then a dentist pulled their fangs and replaced them with normal incisors. She also had them sterilized to prevent them from reproducing, but they were still able to use their huge reproductive organs. Some still live in New City, enjoying productive lives. The Lord Mayor Bergen and his son Ramon retrained them to become contributing members of the Persirian community.

CHAPTER 3

Usually, Lena sets her coffeemaker to automatic the night before. It was no different on this Sunday night before she retired. On Monday morning, she woke to the smell of her Columbian coffee. She jumped out of bed on this chilly morning, wrapped a robe around her and stumbled barefooted to the kitchen, realizing that she has a slight headache. Typically, Lena uses a teaspoon of brown sugar for her coffee, but this morning she poured it and drank it black and without using sugar, hoping that headache will somehow disappear; it had worked before. While she stood under the hot shower, surprisingly her headache mysteriously disappeared. Was it the coffee or the soothing hot water? She wasn't sure. Whatever it was, it worked. Naked and refreshed, with bouncing hard breasts, she swiftly walked thought the living room and glanced out, slightly pulling the drapes apart, only exposing her face. Lazily, snowflakes drifted onto her balcony, covering the floor. She pressed her lips and nodded knowingly. She saw a young man rushing to his car in the parking lot sliding on the fresh snow. He wore minimum clothing and his body was shivering. At that moment Lena made up her mind; she will wear her snow boots and in her second bag, she will take her black high heel shoes for the office. It took her one half hour to get ready. Under her loose, suit jacket, she carried a Glock 18C. Her car was parked it the parking garage. When she turned the key to start the car, the radio news reporter recited news that Lena had trouble following. However, what she surmised was that in selected places of the world, a strange disease killed thousands of people, especially in third world countries, as well as in parts of Africa and Europe. After hearing partial information of the news, Lena raised her brows and pressed her lips, glad that she had nothing to do with *thi*s situation. She switched the radio to local news and soft music.

The drive to the office took thirty minutes on a good day. Though the roads were slippery, it took her only thirty-five minutes, because traffic was light. Most folks used the excuse that school officials delayed the start of the schools two hours. When she arrived in her office, Lena promptly donned her high heel shoes and walked confidently down the hall to see Martin Hall. He purposely left the office door open and Lena walked right in.

"Good morning, Lena," said Hall, with a subdued voice, raising his brows. By the tone of Hall's voice, Lena immediately knew that what he had to say was going to be difficult for him. He definitely was not going to send her on a vacation, besides she just returned from a great skiing trip.

"Good morning, Marty. What's up?" questioned Lena, standing behind his desk.

"How do you take your coffee?" Lena looked at Marty, creating wrinkles on her forehead and thought, *what the hell is he up to now?*

"This morning; black and one spoon of sugar." Marty rushed to the counter and poured two cups of coffee. Then he took them to his desk. He placed his cup next to him and Lena's cup across his desk and said, "Please sit down." Reluctantly, Lena sat, crossed her legs, reached for the coffee, had a sip and said, "You make good coffee." However, whenever Marty

11

sternly asks one of his employees to sit, he will reprimand them or he will inform them of an unpleasant assignment. *It must be the latter*, thought Lena.

"Thank you. Well, we might as well get to the crux of the matter. By chance, have you heard the latest news?"

"I'm not sure, Marty. I listened to the run of the mill news; the weather, car accidents, fires and break-ins. And there was a story about a crocked politician. On the way to work, I heard on the radio that thousands of people are being killed in different parts of the world. Is that what you might be alluding to?"

"Exactly. The director of the FBI called me last night and told me that the president of Tunisia called him, asking for help."

"What kind of help, Marty?"

"I'm not sure that I fully understood, what he said."

"Why didn't you ask him what he was trying to say, Marty?"

"Frankly, I didn't want to embarrass the director and he, in turn, didn't ask the president about details. I don't think that he fully understood the brunt of the problem either."

"So, do you fellows have any idea what the problem might be?"

"No, we don't. All we know is that a disease is killing off thousands of folks. The strange thing is that isolated individuals are not affected. I suggest that you fly to Tunis and start an investigation — President's orders." Lena folded her hands in front of her and chose her words carefully, "If you have no objections, I would like to take Frank Huston along

12

with me. He is one of the best FBI forensic investigators and he has worked with me previously when we dealt with the Ducazians."

"As always, Lena, you have a free rain. However, *please* be careful. We do not know what is killing off the folks down there."

"I have one question."

"Yes, what is it; but I'm not sure that I can answer it."

"Why is it that it's always us that helping everyone around the world? Just a few years ago, the president called us in to stop world slavery, or at least make a dent in it.

"That's a good question. It must be because you are good at what you do."

"Ha! Flattery will get you nowhere, Marty. But, I could use a raise."

"As of right now you are living on an expense account. This will help you. I will have to think about a raise. I must control my budget and I get a limited amount allocated for raises in my department."

"You managers are all alike and you always have a good answer." Marty shook his head and smiled. He knows that Lena understood the politics of management. She also has a limited budget to work with. Then she said, "Well Marty, if you have nothing else, I might as well get started with my new assignment. How much time do I have to investigate this problem?"

"Good question. Frankly, Lena, I don't know what you will find, what you will have to pursue and where the investigation will lead you." Lena nodded, finished her coffee

13

and left. At the door, she turned and she said, "Wish me luck."

"Good luck, Lena. Make sure that you report to me."

Lena returned to her office, leaving her empty coffee cup on her boss's desk and prepared for tomorrow's trip to Tunis, Tunisia. First, she talked to her secretary to reserve a flight for two on a commercial airliner. Then she called Frank. He was surprised to hear from her and she said, "I'll see you at ten at the airport. Make sure that you bring a change of clothing, particularly casual stuff and do *not* forget your complete forensic kit.

"What's going on, Lena?"

"We'll talk during our flight."

CHAPTER 4

Presently there are no commercial, direct flights from DC to Tunis. Therefore, the secretary scheduled the next best thing; a flight from DC to Dallas and from there to Tunis. The departure from DC was early in the morning. Lena stood by the ticket counter and looked for Frank Huston. He was 10 minutes early, standing by a nearby coffee bar, sipping mocha. She saw him in the distance. He chugged his coffee, threw the empty cup in a garbage can, then he dragged his heavy forensic kit on wheels with his left hand, and he had a medium size suitcase in his right. He tried to wave at Lena but it was difficult to move the suitcase up and down, therefore he nodded. Lena realized what he was doing and she waved at him. He smiled, showing dimples on his cheeks and a full set of white teeth, "Hi." Frank placed his luggage next to the counter and Lena handed him his ticket. They were ready to board the plane and both always carried passports and special clearances for their weapons. An airline employee sounded off, "We are now pre-boarding for the flight to Dallas."

"What the hell is pre-boarding and where are we sitting?" asked Frank, raising his bushy brows. He was in similar situations many times before.

"I don't know what pre-boarding is, but I see that people are slowly moving to the gate," said Lena and continued, "Let's go Frank; we have tickets for First Class. Is that alright with you?"

"Absolutely." They found their seats and Frank pushed his seat back to make room for his 6 foot, 2-inch frame. A flight

15

attendant walked up and asked, "Would you care for anything?"

"I could use some tea — and perhaps toast," said Frank. The flight attendant nodded and looked at Lena and asked, "What about you Miss?"

"I'll have the same."

"Coming right up." She left, turned right and disappeared. Lena and Frank settled down.

"So, what are we doing and where the hell are we going, Lena?"

"I don't know much, Frank. Yesterday, I had one of the shortest briefings with my boss and at the same time, it might be one of the longest cases. All he could tell me that a disease is killing off thousands of folks in different parts of the world. One such area is near Tunis. However, he received similar reports from other placed on earth. He didn't elaborate on that point and I didn't ask questions. I have a notion that we'll have enough to do in the Tunis area and where this will eventually take us. The strange thing is that isolated groups are not affected. Then he told me that I should fly to Tunis and start looking. Actually, those are the president's orders. Obviously, he must know something that we don't know. I believe that he personally knows the Tunisian president from his college days."

"Wow. But why should the FBI be involved in this?"

"I'm not sure. Those are the president's orders. Perhaps this could eventually become a danger to the United States."

"Well, whatever it is we will have to find out about it," said Frank, in deep thought, pressing his lips.

"You are right about that …." The flight attendant brought a tray with two cups of tea, sugar toast and jam. Lena and Frank drank the tea and munched on the toast. After the snack, they stretched out for a short time and relaxed; each was involved in their own thoughts. In Dallas, they changed planes and later the same day, they arrived at the airport in Tunis. It is located east of the city. A tall, very good-looking police officer asked them in perfect English for their passports and their transit visa, "What is your business?" Lena and Frank showed the police officer their visas. Then Lena flashed her FBI badge, "We are on official business."

"Welcome to Tunisia," said the officer. "Are you staying long?"

"We don't know where our investigation will lead us."

"What are you investigating, if I may ask?"

"We will try to find out why many of your people are dying, while others are perfectly healthy." The officer stopped in his tracks and he was going to say something, however he changed his mind. He nodded, "Good luck with your investigation." Then he walked away, while he pulled out his wireless communication device and talked to someone. Lena noticed that, but never thought anything about it. She flagged down a cab, while waiting for their luggage. Fortunately, the driver spoke some English with a French accent.

"Take us to a hotel with a good view," said Lena, thoughtfully.

"I know exactly the place for you. It is the Hotel Abou Nawas Tunis. You will be able to see Lake Tunis from your window. It is perhaps the best hotel in the city and it is safe."

"Alright. Take us there," said Lena, impatiently. Frank looked at the driver, evaluated his apprehensive behavior and

17

asked, "How do *you* feel? Are you alright?"

"Funny that you should ask me that. I feel fine, but my older brother died last week. I'm still trying to get over it."

"How did that happen?"

"He died suddenly. But, I couldn't find out why he died. The police took him away and I believe they cremated him immediately."

"Where did he die? In the city?"

"No, no, not in the city; in El Habibia. It is a town west of here. I am not going there. A friend of mine said that a killer disease is killing many people. His brother also died there, but he arrived too late to see how it happened."

"So, you didn't see how your brother died," asked Frank, pressing his lips.

"No sir, I didn't." Lena and Frank looked at each other, but remained quiet. Then Frank asked, "Is El Habibia a big city?"

"No, not too big. Perhaps a few thousand people. Most of them are poor and live in dilapidated shacks or mud huts," said the driver, nodding. He continued, "They come to Tunis, trying to find jobs. Most of them are homeless and steal anything they can — mostly at night. They are a real burden to us." Frank grabbed the handle in front of him, bent forward and said, "I see. Was your brother sick before?"

"Never. He was healthier than I am. When he died he was only forty-two."

"What did he do for a living? Your brother, I mean?" asked Lena.

"He worked in a restaurant — sometimes in the kitchen
18

and sometimes as waiter. Lately, he had a better job. He was the caretaker for the mosque in the downtown area. The man who had the job before him, died also." Frank nodded, indicating that he appeared to understand the situation.

"Thank you for your help," said Lena, showing sympathy. Ten minutes later, the driver pulled up in front of the hotel and brought the rickety cab to a screeching halt. He helped them with the luggage and Lena paid him. And she gave him a decent tip. The driver thanked them; he was overwhelmed and bowed slightly. Lena booked two adjacent rooms on the first floor; both had dinner and then went to bed after a hectic and unusually long day.

CHAPTER 5

Frank was up at six thirty. He had a restless night. He thought, *what the hell am I doing in Tunis?* Being a top-notch forensic investigator, he had a bad feeling about this place, especially since he does not know the neighborhood. He shook his head, trying to shake off his feelings, stretched and took a cool shower. The shower floor was slippery and the walls were dirty. Frank tiptoed around very carefully to prevent from slipping and falling. Also, very little hot water was left, because someone else next door also took a shower. He was forced to finish off with cold water and at the same time, he began to feel the atmospheric pressure bearing down, promising a hot and humid day. Slowly, he dressed casually, went to the dining room and asked for coffee. Surprisingly, Lena joined him only ten minutes later, "I called you in your room."

"Let me guess. I didn't answer the phone." Lena laughed, "No you didn't. Did you order breakfast yet?"

"No I didn't." While she said that, she simultaneously waved for the waiter. He came over and both ordered an American style breakfast — bacon, eggs and home fries. After breakfast, Lena paid, made a note and then they went across the street to a car rental agency. They got a newer model Mercedes van, with a working air-conditioning system. That in itself was a streak of luck in this part of Tunisia.

"How long are you going to keep the car?" said the attendant.

"We are not sure. Perhaps a couple of days."

"No problem." Frank placed his forensic kit in the back.

20

Since it was already 79 degrees Fahrenheit, Frank turned on the A/C unit and he made sure that the he had a full tank of gasoline. Then he drove to the hotel and he parked the van behind it and locked it. At the front desk, Frank asked for a map of the general area. He checked it and saw that it included the town of El Habibia. The agents found a table in the lobby, sat down and studied the map. The town is about 14 miles west of Tunis. They decided to take highway P7 and then follow the sign pointing south to El Habibia, less than two miles. Driving over paved road, badly in need of repair, some areas covered with fine sand, they saw another illegible sign for the city, pointing straight ahead, flapping back and forth in the breeze, metal squealing. Sand dunes were on the left, created by the constant winds across the rolling desert land. Frank approached the town from the north. Ahead of him, he saw the center circle. Slowly he drove around the circle, looking down the narrow side streets. In the center was an abandoned café. A sign was hanging from the building on one nail. It read, Café El Habibia. When Lena saw that she laughed out loud, "This is the ultimate joke." On his right, a sandstone structure occupied an oversized corner lot. It might have been a warehouse at one time but now it was deserted, except a scrawny dog ran from building to building, sheepishly looking back, tail pulled in and howling. Frank kept the Mercedes locked and his gun on the seat, safety off. Lena held hers. They were ready for anything. A short distance ahead, they saw dilapidated mud buildings on their left, the front door partially open. It also appeared deserted, except that a tall, good-looking, well-dressed western type male, watched them approaching in their van. He looked similar to the

21

officer at the airport — it could be the same man. He walked swiftly and he showed no indication of sickness. Eventually he disappeared in the mud building. A few seconds later, the same man looked through a dirty, broken window and watched them. It was unusual for a van in good condition in this run down town circling around the plaza. Frank drove around the circle one more time. He hoped to strike a conversation with the male, but he seemed to have disappeared in his elusive manner. Then, while Lena looked down one of the side streets, she saw something strange and she said, "Stop the van Frank. Look down there." Frank hit the brakes hard, causing the van to twist on the sandy road creating a cloud of dust. What they saw was implausible, and at the same time, sickening. A middle-aged woman stood in front of a store, dusty merchandize hanging in front, moving in the breeze. An unknown circumstance caused her to slump over. That was what caught Lena's eye.

"Frank, drive a little closer, but please be careful." When they were close enough to see the woman's face, they noticed that it was twisted from fear. She had trouble standing. Then she looked for a place to sit. Next to the store was an insecure old bench. She walked to it and sat, facing the detectives. She tried to straighten and only an extreme amount of effort afforded her to lean back. She raised her hand and looked at it. One of her fingers had blood oozing from it. The woman looked up smiled at the agents and nodded, tying to say, "You see. This is what became of me." They were nervous, they didn't smile back, nor did they talk to her. Now she had blood covering her teeth. There was too much blood in her mouth forcing her to spit it out. Not too long thereafter, she had blood discharging from eyes.

"What the hell kind of sickness does the woman have," asked Lena, overcome by horror and sadness.

"It must be some kind of a disease," guessed Frank, while he reached for his Flip camcorder with a wide-angle lens. In the same breath, he turned it on and pointed it toward the wretched woman.

"No shit, Sherlock. Thank you for your professional observation. This is horrible."

"Yes it is. I think that it could be some kind of new virus. Something that I haven't seen before. Apparently, the virus keeps traveling through her body. Now the woman succumbed to her faith and she didn't appear to have pain."

"It must be a mighty powerful virus and it must attack her nervous system first."

"Right."

The virus, in fact, did travel rapidly, throughout the woman's body, attacking the nervous system first. Then it attacked arteries, veins, muscles and other organs. Next, the surrounding skin tissue visibly changed. It turned gray and then bloody. Patches of skin were separating from her body and falling to the ground. Frank and Lena kept watching the woman. Blood was running down her legs, terminating on the sandy ground and running under the bench. Now widespread and severe focal necrosis set in. The poor woman closed her eyes and dropped her head. She felt no pain, but she was still breathing heavy. It appeared that the virus consumed her nervous system first. Apparently that was lucky for her — a virus that kills, while there is no pain. A few minutes later, only

23

her shiny skeleton was left. And since there were no muscles and tendons to support it, the bones finally collapsed, creating an unnerving sound, like bones hanging from a front door at Halloween. Frank and Lena were speechless. They sat in the van, trying to assimilate what they just saw. Pulling himself together, Frank asked, "What do you want me to do, Lena?"

"You are asking me? You are the forensic scientist. Why do you think that I brought you along?"

"Shit, you are in charge. What should I do?"

"You are putting me on the spot, Frank. How about trying to find out what kind of virus this is." Frank nodded and looked back at his forensic kit. It's big and heavy. Barring a kitchen sink, he has everything packed in there that a modern forensic scientist could possibly carry.

"Let's see what I can do. Lena, you better look around you to see if anyone is coming, while I'm working." He gave the camcorder to Lena, "Keep recording."

"Right. I'm watching," said Lena, holding her Glock tightly in her right hand and the camcorder in her left. Frank squeezed his way to the back of the van. He opened the kit and took out a pair of white overalls. He donned them over his clothes and his head. Then he covered his shoes with protective covers. Next he took out a flat package about two inches by two inches and pulled on a string. The package opened up and it became a plastic bucket, marked Biohazard. He put on white, rubber gloves, grabbed a sample kit and shoved it in one of the overall pockets. He was almost ready to leave the van. His gun was still on the front seat of the van and he almost forgot it. Lena saw that and said, "Don't forget your ray gun, Frank."

24

"Right. Pass it to me, please." Lena reached over and gave Frank the gun. He put it in his gun holster under his left arm.

"Be careful. So far, I haven't seen anyone — front or back. Hurry up and get a sample." Frank opened the backdoor of the van and stepped out. He looked around and he saw that the area was deserted. Then he ran to the remains of the unfortunate woman. He used his sample kit and collected samples of the disease, placed the cotton swabs in his containers and closed it tightly. He was ready to return when he saw the same good-looking, tall man approaching him. He held a gun in his hand. Fortunately, Lena saw him too and she immediately stepped out of the van, camcorder running. Now the man was in between Lena and Frank. He raised his gun and aimed it toward Frank. Lena saw the danger and due to her excellent FBI training, she fired a shot, aiming at the man's back. She must have hit his right shoulder, because the man's gun fell to the ground. The man turned to see who fired at his back. While he turned, Frank fired his ray gun. He wasn't sure if he hit the man. Surprisingly, the man ran away, without missing a stride, and returned to the building from where he came. Frank picked up the man's gun and took it to the van. He checked it and it was an American made S&W, caliber 45. He ripped the overalls off his body and stuffed them into the biohazard container. Then he removed the contaminated shoe covers and his gloves, shoved them into the same container and closed it. He put the man's gun into a plastic bag, being careful not to smudge fingerprints and he safely placed all items in his forensic case.

"Let's get the hell out of here, Frank. Something is

25

going on here and we must find out what it is." Simultaneously, Lena turned off the camcorder.

"I agree." Frank drove back to the hotel and while he did, Lena called Martin Hall. She described their horrible encounter in detail.

"Come home immediately, Lena," said Martin, expecting the worst. "We have to make plans, relying on the new evidence that both of you found." As soon as Lena and Frank returned to the Hotel Abou Nawas Tunis, Lena found out that a private FBI jet will leave for the USA next day at seven a. m.

This evening, they had time to kill and stayed in the hotel where they had a typical Tunisian dinner; Coucous, the national dish of Tunisia; lamb, vegetables and spices are cooked together in a pot and various herbs are added, which gives it a distinctive flavor. They also had a warm beer, then they went for a walk to the Medina, in the center of the city. It is a dense area of alleys and covered passages. The locals engaged active trade; it included goods from leather to tin to the finest cloths. Between the alleys, were tiny craft shops, operated by skilled Tunisians pushing to sell their art objects. Lena and Frank enjoyed watching a silversmith forming a broach. When he was finished with it, he handed it to Lena and she was obliged to purchase it.

"Looks nice," said Frank, smiling.

"The next time I'll know better, however, I do like it." They continued their stroll and came upon four middle-aged men sitting around an old rundown table with a burning light bulb above, dangling from side to side. They played one of their favorite board games. Obviously, they enjoyed themselves, but not without shouting and arguing. Lena looked at her watch and said, "I'm getting tired."

"So am I. Let's go back." On their way back to the hotel, Lena thought that she saw the man that she shot in the back watching them. However, she wasn't sure. *Tomorrow we'll have another boring day,* thought Lena.

CHAPTER 6

Marty Hall called Lena and told her that, they could hitch a ride on a FBI jet on their return trip to DC. The flight was boring, however the FBI pilot made good time. They killed time by drinking coffee and eating nuts. When the agents arrived in DC, they donned their winter jackets and took a cab to their offices. Two days ago, it snowed, but in the meantime, the city workers cleared the streets. When Lena arrived at the office, she immediately went to Martin Hall and reported to him. He was deeply concerned about what she told him, but it was too early for him to form conclusions.

Frank immediately went to the lab with his forensic samples. Two scientists, a man and a woman, were still present and appeared to be drifting toward their locker room. They still wore their lab coats, but it was obvious to Frank that they were ready to quit for the day. Frank handed them his samples that he collected in El Habibia.

"What do you want me to do with this?" asked the woman, dismayed. Frank couldn't believe that she asked him such a brainless question, after he went all the way around the globe to get the samples. However, he kept his cool and said, "Could you please run this through your analyzer and tell me what kind of virus this is? By the way, take all possible necessary precautions. From what we witnessed it's extremely hazardous."

"How do you know that this is a virus?" asked the man, trying to look intelligent.

"So sorry. Perhaps Lena McCabe and I have assumed

28

too much. Please do the test and tell *us* what it is."

"All right." Both scientists went to their analyzer, smeared a small quantity on a glass plate and placed it into the analyzer chamber. They waited five minutes and then looked at the results. They had a puzzled look on their faces, as if they screwed up.

"What's wrong," asked Frank, perplexed.

"We have to run the test again," said the woman, shrugging her shoulders. Five minutes later, the printer spit out the results again, with such force that the paper drifted to the floor. The woman picked it up, looked at it and then handed the results to her partner. "Yep, the same result," said the man. In the meantime, she reached for an oversized book, listing viruses, flipped through it, read a paragraph and handed it to her partner. "Yes, that's what I thought." Frank was getting impatient.

"Where did you get this sample, Frank?" asked the man, ignoring Frank state of mind.

"I got this sample in El Habibia." Both forensic scientists pressed their lips, shook their heads and the woman asked, "Could you please give us a straight question. Where the hell is El Habibia?"

"It's west of Tunis. That's in Tunisia."

"We know where Tunis is. Why are you so reluctant to tell us all the information that you have?"

"Because I want you to form a finite, independent conclusion about the sample and not about the surrounding circumstances," said Frank, impatiently.

"That makes sense," said the woman, raising her over

29

plucked, thin brows that looked like semicolons. She thought for a while and continued, "You won't like what I am going to tell you, Frank."

"Tell me already."

"Well, our tests show that this is a new strain of the Yambuku, Ebola virus. Bacteriologists classified this strain as the Yambuku, DC virus. The original virus had an 88 percent mortality rate and an incubation period of about two days. This new, DC, strain incubates in three hours and it is 100 percent deadly. Moreover, no one found a cure for it. Agents, I wish I could say that our analyzer is defective, but we had it recalibrated just one week ago, per our ISO 9001 quality control system. You can check the certification tag, if you like." Frank sat on a nearby chair and said, "I'm pretty sure that neither you nor I are infected. Because if we were, all of us would be dead by now. I can vouch for that. I have seen this virus at work."

"Absolutely."

"Let me ask you another question. Is there any method to kill this virus?"

"Most definitely. You have to disinfect yourself and everything around you with a 0.5 percent sodium hypochlorite solution. Or if possible, burn everything around you."

"Well, that's encouraging," said Frank, standing up. By the way, where does this Ebola virus originate?"

"It originates in Zaire, near the Ebola River," said the woman, "and both of us are wondering how in the hell the virus traveled from Zaire to Tunisia, though both are on the African continent?"

"That's a good question. I believe that Lena and I will

30

have to figure that out. By the way, where can I buy this hypochlorite solution?" asked Frank.

"That won't be necessary. You *cannot* buy this on the open market. We'll give you a bottle." Frank nodded and asked, "What are you going to do with my samples."

"We are confiscating them and we will keep them in one of our cryogenic units. I will also make an entry in our database. That's important evidence and it will become a permanent record in out FBI files" said the male scientist.

Frank thanked the scientists for their information and rushed to Lena's office with a copy of the report, hoping to catch her still at work.

"Hi. I'm back. How did it go with Martin?"

"Same as always. And we are waiting for your results."

"I got them, Lena." Frank related to Lena the findings of the scientists and gave her the report. She was stunned and she and called Martin immediately.

"Both of you. Come to my office." Lena and Frank went to Martin's office down the hallway.

"Coffee anyone?"

"No, thanks Marty. Not after all this." Lena placed a copy of the scientists' results in front of Marty. Marty studied it for a long time. Actually, it didn't take him long to read it. While he looked at the paper, he tried to formulate a game plan. At last, he looked up at Lena and asked, "What do you think that we should do?"

"I'm not sure, Marty. However, I do have questions."

"Like what?"

"Well, I would like to know who the man was that tried

31

to stop us, while Frank was collecting samples from the skeleton of the dead female body. By the way, we have a video for you."

"Let me see." Lena gave Marty the Flip camcorder and he witnessed the death of the woman as well as the well-dressed man running toward the building.

"Good job," said Marty. "I understand that you shot him in the shoulder."

"Yes, I did. I must have severed a tendon or various tendons, because he couldn't hold his gun and he let it drop to the ground." Marty looked at Frank, "Didn't you shoot him too?"

"Yes, I did. I must have missed him because he didn't seem to be affected by the force of the ray. Anyway, that was what I concluded."

"Honestly, Frank I don't believe that you missed — not with your training. Perhaps he was wearing a bulletproof vest," said Marty, raising his brows.

"If he was wearing one, his body should have undergone a jerking motion — keeled over or twisted sideways, considering the powerful impact force," added Lena. "But none of the above happened."

"I agree," said Marty, still holding the Flip camcorder.

"So, what are we saying?" asked Frank, frustrated. "Are we saying that this person was superman?"

"Who the hell knows," said Marty. "However, I know one thing for sure. Now that you know what kills those people, you have to go back to Tunis and try to find out who this man is, where he is coming from and what his movements are.

32

Secondly, how the hell does the Ebola virus get from Zaire to Tunisia? I also have to tell you that I have had word that folks are getting killed off in other parts of the world." Lena shook her head, realizing that she might have to take other trips. Then she looked at the watch, "It's getting late. I would like to leave on Monday morning to return to Tunis and I would like to take Julius with me. If that's acceptable."

"That's fine. So, you won't need Frank for this trip?" asked Marty.

"No. If I'll need him again, I'll call for him. He did a super job. Taking him on this trip might be overkill," said Lena, looking at Frank.

"I go along with that," said Frank, smiling and continued, "What do you want me to do with the stranger's gun?"

"Have it checked for finger prints. Then call me on a wireless or send me a report," said Lena. Martin Hall stood, signifying that the meeting is finally over, "Have a nice weekend, both of you."

"Same here." Both returned to their offices, organized their notes and finally went home.

33

CHAPTER 7

It's been eight days since Marty Hall gave Lena McCabe her new assignment. It's early Monday morning and on this trip, she is on a FBI jet, again headed to Tunis. This time she went with her longtime associate, Julius Hersey. They work well together and Lena decided that both should carry their ray guns, rather than old-fashioned noisy guns, using steel covered projectiles. The new weapons are much quieter, lighter and they are extremely effective. From her last encounter with a stranger, she feels that their ray guns might be more efficient than the old-fashioned Glock handguns against the man who didn't seem to be affected by bullets. They carried other weapons on the jet, but they will only use them if circumstances force them into it.

Lena had ample time to relate to Julius their first run in with the strange man, while she showed him a copy of the video. When she thought that she finished Julius asked, "Are you sure that you hit the man?"

"Positively. You know that I never miss at this short distance," said Lena, frowning. "And the same goes for Frank Huston."

"I have to agree. I have seen both of you on the practice range." Lena pressed her lips, nodded and smiled.

"Did he carry a bullet proof vest or some other form of protection?"

"I don't know, Julius. I couldn't make that out. He wore a loose jacket," said Lena, leaning back in her seat. Julius looked at her and decided to stop asking questions. Lena sat back and tried to relax when her wireless vibrated. She retrieved

34

it and saw that Frank Huston called her, "Yes, Frank. What do you have?"

"I had a lab technician check the stranger's gun for finger prints."

"What did the technician find?"

"Nothing. He couldn't find finger prints, Lena. I thought about it. The stranger didn't wear gloves and he didn't have time to wipe off the finger prints."

"You are absolutely right. This case is getting more complicated by the minute."

"Right you are. I have to think about that. Thanks for the info, Frank." Lena disconnected and turned to Julius. She related the conversation to him. He shook his head, leaned back in deep thought, held her hand and both fell asleep. It's a long flight from DC to Tunis, about 4,500 miles. With the FBI jet, at approximately Mach 1, they flew nonstop for a little over six hours. They also reviewed all the information that they had at their disposal. They studied Tunis and before they arrived, they ate on the jet. Lena flagged down a cab and she directed the driver to the Hotel Abou Nawas Tunis, the same hotel that she used with Frank Huston. This time however, Lena and Julius stayed in the same room. They showered, donned comfortable clothes, hiding their ray guns, and went for a walk in the Medina area. They kept their eyes open for any irregularities or strange looking men that didn't fit the norm. Lena also had her Flip camcorder in one of her side pockets. They walked through a maze of tunnels and alleys dotted with hidden mansions and in between, they visited hole-in-the wall workshops, with sun beaten and elderly men performing their crafts. Most of them

35

looked bored, hoping for a sale. Then the narrow alley opened up to a glittering, modern shopping mall, filled with Arabic folks, dressed-to-kill; many were dressed in local garb. Lena and Julius stopped in one of the coffee houses, adorned with giant keyhole shaped doors. They sat, ordered coffee and pastry, they enjoyed themselves and they observed the surroundings. Criminals loved Tunis, because the narrow alleys are excellent places to hide, but today neither detected unusual movements. As they finished their pastries, they paid and Lena drove their rented Ford, to El Habibia, the place where she saw the woman dying. She parked and guided Julius to the same side street, where she and Frank had their encounter. The street was deserted and no one walked the walk of death. Lena thought, *did I dream all this the last time I was here?* Julius had similar feelings and said, "I know that I saw your video, but are you sure that this was the street where you saw a person bleed to death and collapsed, with only the skeleton left?"

"Yes, Julius. That is the street where Frank and I saw that." The old bench was still standing in front of the mud building. Then Lena turned and pointed impatiently at a long building where she saw the stranger disappear, "This is where the stranger went in."

"Did you follow him?"

"No we didn't. After all Frank is a scientist, not a FBI agent, trained in the art of killing, though he is an expert using a gun."

"I understand," said Julius. "Let's go and check out the building."

"Fine. I hope that you brought your flashlight."

36

"I never go anywhere without it. How about you? Do you have your ray gun?"

"Absolutely." They left the Ford where she parked it. Both slowly crossed the street and looked up at the building and at all the windows. A grubby, old dog, which must have lost his master, ran down the street, past the building. He looked back sheepishly and continued on his track.

"I think that I saw the same dog, the last time that I was here with Frank," said Lena, frowning. Julius nodded. He reached the front door first, grabbed the door handle and opened it.

"It's not locked."

"Yes, I can see that." Julius looked at Lena, wondering if she is humoring him. When he looked at her, she smiled and shrugged her shoulders. They entered a long, narrow and dark hallway. On the other end was an exit, perhaps another entrance — old country construction.

"I bet this is where the man escaped when he saw us last week."

"Very possible; unless he lives here."

"Who the hell would want to live *here*, Julius?"

"Perhaps a local with limited means."

"That's possible. But the man I saw last week didn't appear to be with limited means." Julius opened the first door on the right. It was a two-room apartment, where someone appeared to have left in a hurry. Dirty dished were still in the black sink, flies sitting on the dishes, sucking up what's left of the dry food specs; brown water dripping from the faucet. The bed was unmade; bed sheets were missing and the mattress had

37

bloodstains on it. The toilet had dried up — feces stuck to the bowl.

"What could have happed here?" asked Julius, confused.

"I believe the occupants must have left in a hurry. Look at the floor. Dry blood just like on the mattress."

"Yea, I can see that. I'm leaving, before I throw up." Julius turned and walked toward the door. Lena followed. Carefully and quietly, they walked halfway down the hallway. Julius opened another door. It was also unlocked. Similar conditions existed there and more blood was visible not only on the floor but also on a chair. A small-screen TV, in black and white showed an Arabic, veiled woman talking. They shrugged and continued. Now Lena anxiously took the lead. She opened the last door on the left. To their surprise, this room was immaculate, minimum furniture but a large TV on one of the tables. Immediately they saw a Caucasian-type man standing near a table; a large birdcage was on top. He was well dressed, about six foot three and very good looking.

"I think this is the man that I saw before. He is wearing similar clothes," whispered Lena.

"I agree."

"Don't you know that you should knock before entering a private dwelling?" said the man, defensively with an English accent.

"We are sorry, but we didn't expect that anyone would live here. The place seemed deserted," said Lena, prying.

"I am staying here. Can't you see that?"

"Yes, we can see that. What's your name?"

"My name is David Barkley. What can I do for?"
38

Barkley fumbled with the lock of the birdcage.

"What do you keep in there," asked Lena, raising her brows.

"They are bats. Any objections?"

"No, no. Definitely no objections. They don't seem to be the kind you have in North Africa."

"They are not. Originally they are from Zaire."

"I see," said Lena. She looked alarmingly at Julius, and squeezed his hand. He understood and said, "Well, we are sorry to intrude. It's late and we must be going. He held Lena's hand and dragged her to the door.

"It was a pleasure meeting you, Mr. Barkley."

"Knock, the next time," said Barkley with a controlled baritone voice. Then he turned and continued with his business, but neither Lena not Julius knew what it was. Julius opened the door carefully and he said, frowning, "He is a strange person." When the agents returned to their car, Julius drove as quickly as he could to their hotel. After a long day, both were tired and hungry.

CHAPTER 8

A waiter guided Lena and Julius to a table, near the window. It was covered with a red tablecloth and that is the color of appetite in Tunisia. Looking out they could see the plaza. On the far end was a mosque, with people slowly entering and leaving, still busy this late in the day. Before they had an opportunity to engage in a discussion about the mosque, a waiter passed a menu and said, "When you are ready I will take your order." Lena looked at the menu; she had difficulty deciding what to order and she called the waiter, "What would you recommend?"

"Today we have one of our specialties. It is coucsous. It is a stew, supported by soup, salad and vegetables. The chef just barely finished cooking it." Lena looked at Julius; he nodded.

"Alright. We will have your specialty."

"I will also bring you a pitcher of ice water, because our foods are pretty hot. Would this be alright, ma'am?"

"Yes, that's alright," said Lena, expecting approval from Julius, since he left her with making decisions, concerning their food. While they waited for their dinner, Lena pulled her minicomputer from her purse. Julius saw that and asked, "What are you up to?"

"I am going to do research on bats, Julius. We'll see what they have stored on the FBI database."

"What kind of research?"

"I will tell you if you give me a chance and as soon as I find out."

"Alright! It's another working dinner."

40

"You might call it that. I want to find out why Barkley has bats from Zaire in El Habibia."

"Interesting. I also wondered about that." Lena nodded and hooked up with the satellite system. Quickly she found out a very disturbing fact. When Julius saw the look on Lena's face, he became worried, "What's wrong, Lena?"

"You won't believe what I just found out."

"What? Tell me! I can hardly wait."

"The Zaire bats are carriers of the Yambuku, Ebola virus. Bacteriologists classified this strain as the Yambuku, DC virus."

While the waiter placed the dinner on the table, Julius could not help himself and blurted, "Son of a bitch. Who would have thought that this was possible?" The waiter nodded, smiled and said, "Enjoy your dinner and please, don't forget to drink your water." Then he left, concerned that Julius was talking about the food. Both agents were in deep thought, troubled about the repercussions of the disease carrying bats. They continued slowly eating their dinner.

"Wow!" Julius reached for his glass of water and drank all of it. Then he poured himself another glass and asked. "Do you need more water, Lena?"

"No thank you. I love hot foods. I grew up with them. My mother frequently made Mexican dishes." Both had fruit for dessert and then looked at each other. Both started their conversation, "What should we do next?" Despite their serious concerns, they laughed. Julius continued, "Let's drive back to the apartment building and find out more about this David Barkley."

41

"I agree." Lena paid for the dinner, left a normal tip and charged it to her expense account. Then they left and Lena drove to El Habibia. A strong wind blew from the west and deposited sand on the road, making it difficult to stay on the road without sliding. When they arrived. it was late afternoon. She parked the Ford on the blind side of the apartment building. Both walked toward the building. This time they entered on the other end. The top of the building was still illuminated by the slowly sinking sun and they noticed that the temperature dropped quickly. A desert critter ran from the building and disappeared in the distance. Lena pulled her ray gun and armed it, flipping the safety latch to OFF, and she zipped up her jacket. Julius looked at Lena, pressed his lips, nodded, and he did the same. Julius opened the front door and to his surprise, it was still unlocked.

"Don't you think that Barkley might have locked the door, after he had unexpected visitors. This appears to be strange behavior."

"Perhaps it's not his apartment building."

"You are right. Let's see if his apartment door is also unlocked." Again, Julius took the lead. Quietly he opened the door to Barkley's apartment. He looked at Lena and pressed his lips. Both understood each other. When they looked in, Barkley stood near the wall, facing the table and he didn't move. The two FBI agents slowly walked closer to Barkley, ray guns in front of them. Barkley still didn't move. Finally, Julius said, "Mr. Barkley, are you alright?" Finally, Barkley slowly turned his head and looked at the intruders. His body shook and his eyes glowed for a moment, then he responded, "Did I not tell you that you should knock before entering?" Julius took a chance and lied, "We did knock but you didn't answer." Barkley's head moved from side to side and then answered, "Sorry, I did not hear you. What can I do for you?"

"We are FBI agents from the United States of America and we are going to confiscate your bats." At the same time, both Lena and Julius approached the table in the middle of the room. When Barkley saw that, he rushed for the table and picked up the birdcage with the bats inside.

"They are my bats. You cannot have them," said Barkley, with a hoarse voice, eyes flickering.

"Let go of the cage, Mr. Barkley. As of now, these bats are United States property," said Lena, pointing her ray gun at Barkley. He did not follow Lena's instruction, turned and ran toward the apartment door, holding the cage. Julius pointed his gun at Barclay's back and fired. Barkley stopped in his tracks and stood still. Both Lena and Julius could not understand this strange behavior. Normally, when a person receives a full charge with a ray gun, he collapses and has a hole in his body. Most people would also scream. Yet, Barkley just stood there, not moving. Julius walked up to him, grabbed the birdcage and pushed him. Barkley's eyes had a strange sheen to it and he finally fell backwards, hitting the floor as if an oak tree toppled to the ground. Lena noticed that the skin tore on Barkley's elbow and she said, "Look Julius, he is not bleeding."

"You are right. That is strange," said Julius, perplexed. He placed the cage next to him on the floor, bent down and checked Barkley's elbow. A large gash exposed his elbow joint and he said, "Take a look at this, Lena." Lena studied it and she

43

concluded, "Barkley has an artificial elbow joint." Then she tried to turn the body to check the back but it was too heavy.

"Can you help me Julius?"

"Sure, why not." Now Julius reached down and tried to flip the body.

"I can't do it. Why would the body be so heavy? It must be at least 600 pounds."

"I have an idea," said Lena, and proceeded to open Barkley's jacket and shirt. What Julius and Lena saw was not what they expected. They expected to see blood on his chest. But what they saw was that Barkley has a thin metal plate mounted in front of his chest. Four stainless steel screws held the plate in place. Julius reached in his pocket and removed his all-purpose knife, unlocked the screwdriver and removed the screws. Then he removed the plate. At first, Lena and Julius did not say anything. Then Julius blurted, "It's a fucking roboid. It's the best model that I have seen."

"Right. Look at the hole in his back. It's from your ray gun. Now I know why Barkley remained standing. You severed his main cable, leading to his control center and branching to part of his lower extremities." In the meantime, Julius read the engraving on the chest plate:

SN: C 86,592
Date of Mfg: 5/18/2014
Place: Dallas, TX
Roboid Name: Arthur
Human Name: Charles Barkley
Program: Universal, Population Control
Languages: English, German, Arabic, Italian

Then Julius stuck the plate in his pocket.

"Population Control," said Lena," what does that mean?" In one of Julius closed FBI meetings, they talked about population control and it took him ten seconds to answer, "I'll be a son of a bitch. I never thought that I might get involved with *this*."

"Get involved with what, Julius?"

"It seems that roboids are attempting to control the population of earth."

"How?"

"By releasing bats in predetermined parts of the world. They are carriers of the Ebola virus and when they make contact with a human, they will transfer the virus and he will quickly die. Then anyone that touches the remains will also be affected." Lena looked at the birdcage on the floor and said, "What are we going to do with the cage?"

"We have to get a heavy plastic bag and place the gage in it then stitch it together."

"Right. Let's do that and take it to our jet and take it to the FBI lab."

"What should we do with the roboid?" asked Lena again.

"Good question. I don't know, Lena. What should we do?" Both thought before Lena had an idea, "I think that we should call the Police chief. He might know what to do."

"Right." Lena called the police department of Tunis. An English-speaking woman transferred her to the police chief. He was resting in his office and he said in fluent English, "Yes, this police Chief Christopher Barony. What can I do for you?" Lena

45

explained that she has Charles Barkley incapacitated and that she needed it moved to the FBI jet at the airport. Lena waited for a response and finally Barony said, "Please leave the body where it is. I will be right over and I will take care of it."

"Thank you, Chief." Lena closed her cell phone and told Julius what Barony said. She appeared satisfied.

"Interesting. Should we wait for him?"

"Let's go outside and look around. Perhaps we can find another individual that doesn't fit the norm."

"I'm with you." Both agents left the apartment building and walked across the street, looking at different side streets. By chance, Lena looked back and saw a man running extremely fast into the apartment building. Both returned to the apartment building. The door slammed and before they got near the building, a tall man, about 6 foot, 4 inches carried the roboid from the building. He also carried the plastic bag with the birdcage inside. As soon as he cleared the apartment door, the man ran faster in the opposite direction, then anyone would drive a car.

"How the hell can anyone carry 600 pound and run so fast?" said Julius.

"Julius, I believe that I can answer that question. He also was a roboid. It's too bad that we couldn't get a better look at him."

"Should we call the police chief again?"

"Forget it. He will deny everything — perhaps the police chief is also a roboid. The next time we'll have to be smarter."

"Right. However, I have learned one important fact."

46

"What's that, Lena?"

"Roboids are infiltrating Tunis from the top down and most Tunisians don't know it." Julius and Lena slowly returned to their car and returned to Tunis.

CHAPTER 9

After a demanding Monday, Lena and Julius tried to relax at the hotel. They had dinner and then they returned to their FBI Jet. While driving to the airport, Lena looked constantly behind her. Julius noticed it and he asked, "What are you looking for, bats?"

"Exactly," said Lena, deeply concerned and continued, "You don't know how lucky we were that we didn't have contact with one of Barkley's bats or that we touched an infected item. We could be dead by now, nothing left but our shiny, white skeletons." Julius Hersey shivered at the thought, "We better take protective clothing the next time we chase roboids in a critical area."

"Right. And bats," said Lena, deeply troubled.

"How can we kill bats?"

"We can't. They are under protection. If we kill them, both of us might face a stiff fine. For now, all you can do is isolate them."

"But they are carrying the Yambuku, DC virus."

"I know. Perhaps someday the Department of Wildlife might make an exception." Both laughed.

One hour later, both agents boarded the jet and they immediately went to sleep on their pull-out beds. After they arrived in DC, the pilot let them sleep. Lena woke at seven, showered and had toast and coffee on the plane. While munching on her second piece of toast, Julius said, "Why didn't you wake me."

"There was no need. Get ready and then we can drive to

48

the office from here."

"Good Idea. That will save some time," agreed Julius, raising his brows.

Martin Hall was impatiently waiting for Lena, while Julius sat across from Martin. A couple of minutes later, Lena walked in, a glass of water in her hand, "Sorry that I made you wait."

"That's alright," said Martin. "So, what did you find out?" Lena looked at Julius, raising her brows.

"Go ahead, Lena."

"Alright, here it goes. We believe that roboids are slowly attempting to take over the world by deliberately killing off non-productive humans. At least, that's what we found in El Habibia. This is a small city east of Tunis. Furthermore, we believe that roboids are infiltrating selected areas in the world from the top down."

"How do you know all that?" said Martin, frowning. He carefully considered the next statement, "You have only been down there for a very short time."

"We were there long enough, Martin. After we put David Barkley out of action, we tried to move him but he was too heavy. I asked the police chief of Tunis for help. He said that he would be right there. Then we took a short walk. A few minutes later, we saw a man that looked just like Barkley running away from the building, carrying Barkley as well as the birdcage with bats in it. He ran away from us; as fast a car could drive. We had no chance of catching him."

"How do you know that the police chief carried Barkley?"

49

"It was pretty obvious, Martin. Only the police chief knew where Barkley was. However, it might have been possible that he sends one of his henchmen. Actually, it would have to be another roboid. I know of no one else in this world that could carry a 600-pound man and run at the same time." Martin sat back in his swivel chair. He wasn't ready to give another order, particularly considering that his boss and the president were watching over the case and breathing down his neck. After the meeting with Lena and Julius, he will call his boss and he, in turn, will call the president with the information that Marty obtained. He knows that they knew more that they are telling.

"Is there anything else that you can tell me about what happened in Tunis, or in the town near Tunis? asked Marty. By the way, Julius write down the name of this town." While Julius wrote, Lena continued, "Yes, there is something else that you might be interested in, Martin."

"What is that?" Julius reached in his pocket removed the metal plate and turned it over to Martin and he said, "I removed this plate from the roboid's chest." Martin read from the plate and looked at it for a long time. Past and present events flashed through his mind and it opened a completely new chain of possibilities.

"I didn't know that someone in Dallas, Texas is building roboids."

"We didn't know that either, Marty." Martin continued, "Do you remember a man by the name of Charles Bergen?"

"Yes. He was and still is the Lord Mayor of the Annunnaki Persires Space City, in the area of the Great Basin Desert," said Julius.

50

"Exactly. And as far as I know, he was the *only* one that was allowed to build roboids. Am I right?"

"You are absolutely right," said Lena.

"So, who the hell gave the people in Dallas permission to build roboids down there?"

"I don't know," said Julius. Then another thought came to him and he continued, "There was one incident in their factory, as I remember. That was when Ramon Bergen, the son of Charles, ran the factory."

"What incident?" asked Martin.

"Boy, it has been a while. I'm trying to remember … Oh, yes. A roboid by the name of Sergeant Phebious was missing. Do you guys remember that?" asked Julius.

"I sure do," said Martin, continuing, "I believe that they never found him."

"You are so right. Do you suppose that this Phebious started a roboid factory down there?" said Martin.

"It's possible," said Lena. For quite a while, no one talked. Then Martin Hall stood and walked to his window. A few snowflakes dropped and others circled in front of his window. He turned, faced Lena and Julius and said, "Now I know what you have to do. Take the FBI jet to Dallas and find the roboid factory down there." Lena and Julius stood and were ready to head out.

"One more thing, guys," said Martin Hall. The two agents turned and expecting the worst.

"What is it now?" asked Lena, concerned.

"You should take a break today and leave tomorrow morning."

51

"Thank you, Martin." Hall smiled at them and waved them off. Both looked at each other and they knew where they are ultimately headed — to Lena's apartment. First they went for a stroll in the mall, studying new equipment in a sports store. Then they went to a department store, where Lena bought a pair of black shoes and Julius bought a tie. Next, they stopped at a Chinese, take out restaurant and ate in. Lena had a bottle of wine left over from the holidays and they enjoyed it sitting by her forced air gas fireplace.

CHAPTER 10

Julius got up and stretched. He walked barefooted to the balcony and looked out. A couple of people hurriedly walked across the street, collars turned up. They boarded a transit system bus. He thought, *I had better use my heavier overcoat today.* He went to the bathroom, shaved and took a shower. Fortunately, Julius had an ample supply of spare clothing at Lena's apartment and dressed. Then he made breakfast for Lena and himself — scrambled eggs and home fries. While they sat by the table, finishing their coffee, both were in deep thought. Finally, Lena started, "I know what you are thinking."

"Really. I am thinking that I had a wonderful time last night."

"Men. Is that all you think about?"

"Yes, most of the time. Don't tell me that you didn't enjoy it, after all the moaning and groaning that you did."

"Yes, I enjoyed every minute of it. But, now I am thinking about Dallas. How are we going to find out where the roboid factory is?"

"Good question. Perhaps we could tap into the FBI files and find something that might connect us to roboids."

"Good idea," said Lena. "Let's do that on the plane. One hour later, they were on the plane, heading toward the Dallas airport. Two agents that neither knew hitched a ride. They engaged in small talk, however they quickly settled to do their business. Lena made several entries on her laptop and finally she hit on 'Roboid Repair shop'.

"There is a roboid repair shop south of Dallas, with an

interesting write up but I don't see a factory. A fellow by the name of Paul Marino appears to be the owner. He has his own web page where he advertises Roboid Repair. That's all that I could find," said Lena, excitedly.

"This is more than I expected. Let's go there. Did you find an address?

"Of course, Julius. He is advertising his business."

"Lena closed her laptop. One of the other agents went to the head, pressed his lips, passing her up. Just before noon, they arrived in Dallas. They walked to a car rental agency and Lena rented a newer model electric powered Ford.

"Is the engine running?" asked Julius.

"Of course the engine is running." Lena looked at Julius in disbelieve and shook her head. Sometimes Lena is too serious and she doesn't know when Julius is joking. She entered the address of the roboid repair shop and immediately, a friendly computer-synthesized voice started to give directions. Thirty minutes later, they arrived at their destination. It was outside the city limits. Lena parked the Ford in a parking lot, reserved for customers. On their left was a two-story, ranch type home and on their right was a whitewashed brick building with a red, simple sign adorning the front:

ROBOID REPAIR
Paul Marino, Owner

Behind it was another brick building, with a garage type door, presently closed.

"I guess, that's it," said Lena. "Let's go to the store. Both started to walk. They looked beyond the home and they saw an immense meadow, with an abundance of grazing horses.

"Don't tell me that this is a farm," blurted Julius.

"Let's find out." Julius opened the door, leading to the store. A good-looking, tall man in his forties greeted them, "I'm Paul Marino. How can I help you?" Lena and Julius introduced themselves as business people.

"We are looking for a female roboid trained for domestic duties."

"I am sorry, but we do not manufacture roboids. We just repair them."

"We thought that you might have an older one. Perhaps a trade in?" asked Julius, hoping for a lead or a slip by Marino.

"We do not engage in second hand merchandise. We repair faulty parts and on occasion we reprogram them for different specialties," said Marino, slightly perturbed.

"I see. On the way in, we noticed that you have beautiful horses grazing in the pasture."

"Yes, we do have that. I purchased this farm, including the horses. It is a good income for us."

"That's smart. I presume that you sell off your horses," said Lena.

"Yes, we do that. That is how we invested in capital for the repair shop."

"Can we see your repair shop?" pried Julius. Marino's eyes flickered and then he said, "I suppose that I can show you our shop. Please follow me." On the way to the shop, Lena asked about the building next to the shop, "What do you keep in there?"

"This is where we do our animal husbandry. You know, it is the latest in technology. However, the basics haven't

55

changed much, if you know what I mean. Would you like to see it and I could explain it to you?"

"No, thank you," said Lena. She had no interest in watching horses copulating. Marino was still in the lead. He opened the door to the repair shop. He stopped and as soon as they were inside, Marino said, "This is as far as I am allowing you to go. As you can see, right now my people are working on three roboids. This is a slow week. Is there anything else?"

"No. Thank you for the tour," said Lena sarcastically. She reached in her pocket and grabbed a stickpin. As they turned, Lena, by intentional mistake, got too close to Marino and held on to his arm. While she did that, she pushed her stickpin into Marino's arm. As she expected, Marino didn't react to the pain. While returning to the car, Lena said, "Can we look at your horses? They are beautiful."

"Be my guest. I must return to the store. When you are done, you can leave." Julius was confused. He had no idea what Lena did. Quietly, she explained to him about the stickpin.

"You have guts, Lena," following Lena's lead.

"I know. Let's look at the horses," whispered Lena, realizing that Marino must have turned up his sound receivers, trying to hear what they are talking about. They walked to one of the horses and stroked his neck.

"Isn't this a beautiful horse?" said Lena in normal tone.

"Do you like it?" asked a female voice. Both turned, surprised that they heard a female voice. They saw a tall and beautiful female standing behind them. Somehow, she didn't fit into a farm environment, walking with her high heel shoes.

"Who are you?" said Lena.

56

"I'm Lexia Marino. Antony is my husband." Julius looked at Lena and he raised his brows; perhaps she knows something and he introduced himself and Lena. Then he said, "We were looking for a new or a second hand robot. Do you know where I can purchase one?"

"Unfortunately, we only repair them. However, there is a place called New City. A fellow by the name of Ramon Bergen runs roboid manufacturing plant. Perhaps you could try to purchase one there. They are of excellent quantity. I know, because my husband repaired a few."

"Very interesting. Thank you for the information. We might try that. By the way, what's in the building over there? It seems unique." Lexia seemed to clam up, but eventually came up with an answer, "Animal husbandry. Yes. That's what the building is. My husband performs animal husbandry in there." Lexia turned, unsure of herself, "I must go now. It was a pleasure meeting you." Julius waved at her and she returned the greeting. Then Lena looked into the near distance to the pasture. About fifty feet away, was an island of shrubbery, with fine, greenish-yellow leaves.

"Look at the leaves over there, Julius. They are oscillating as if there is a breeze, but there is no breeze," whispered Lena.

"Yes, yes. I can see that. What do you suppose this is?"

"I think that this is a ventilation system," whispered Lena. "Let's get the hell out of here."

They stepped into their car to return to the Dallas airport.

57

CHAPTER 11

Julius drove to the airport. Contrary to the weather in DC, it was beautiful and the temperature was 72 degrees. He pulled the car into the rental agency and Lena paid for the car rental. At the airport, they had cheeseburgers and small cokes. Earlier, Lena called ahead to tell the pilot when they will arrive for the return trip to DC.

The pilot had the engines running when they boarded the jet. Lena and Julius sat back in their comfortable seats, returning on the same day to DC. Their minds were working overtime and both waited for the other to start a conversation, related to the case that they are working on, while the pilot flew the jet high above cumulus clouds. With the exception of the rhythmic humming of the three jet engines, it was quiet, except that the pilot played soothing, soft music. Lena looked to the left, Julius looked to the right, and they smiled at each other.

"Are you staying with me tonight?" asked Lena, prying.

"If you want me to. I have no urgent business at my place."

"Alright. Then it's all set." Lena rested her head on Julius' shoulder and nearly fell asleep.

"So, what's our next move, Lena?"

"I'm not sure, but I'm still thinking about the foliage moving nervously in the absence of a breeze," said Lena, straightening out in her seat.

"So am I. We determined that this could be the exhaust of a ventilation system."

"Exactly. I bet that he has an underground complex; that

58

would be the only reason that Marino would have a ventilation system in an odd place like this."

"I agree. We need to go down there and see what Marino has down there," said Julius, pondering.

"How? It seems nearly impossible." Lena stared at her cell phone, watching a ten-second commercial, with a caricature jumping in and out of a backyards swimming pool. Then she continued, "I have an idea, Julius."

"What kind of an idea?"

"What do you say, we go to Ramon Bergen and buy a roboid. Program him with a failsafe, protected program and a cover story for our needs and get him to apply for a job at the Marino, roboid repair shop. Perhaps he could find out what they are doing underground."

"I knew it! Leave it up to Lena. She will come up with something original."

"Thank you. I call it, fighting fire with fire." Julius raised his brows and smiled. Then he turned serious and Lena noticed, "Now what are you thinking about?"

"What are you? A mind reader?"

"Sometimes I wish I would be. Perhaps it's best for us that we never progressed to this point. So, what's on your mind, Julius?" Julius leaned back in his seat and started, "I'm worried about the future. Lena."

"How's that?"

"I'm thinking about the roboids. Will they be able to reprogram themselves in the future? And might they have enough impetus to start taking over towns and eventually the world?"

59

"Julius, I think that they are already taking over towns. Just look at Tunis. They have a Roboid police chief and no one in town knows about it. And the town that we visited is practically annihilated and no one asks questions."

"You are right. In addition, I think that they will clean up nonproductive areas by killing the folks and start new in metropolitan areas, keeping humans as workers."

"Boy, that's a scary notion. Don't you think?"

"Yea." Julius crossed his arms in front of him and looked out the port hole of the jet, "We are almost home; just crossing into Virginia. It's a beautiful state."

"Commonwealth, Julius."

"Right. But the university is called Virginia State University." Lena smiled, "You are getting too smart for me, Julius."

"I don't think so." He rested his head on the pillow behind him, "Lena, I have a feeling that this not the end of this project, but just the beginning."

"You are so righty."

CHAPTER 12

When Julius and Lena arrived at her apartment in DC, they ordered in Chinese. While they waited for their dinner, Julius called Charles Bergen, the Lord Mayor of New City.

"It is so nice to hear from you again, Julius. But, I am in the middle of a meeting," said Charles, "What can I do for you?"

"We are interested in buying one of your roboids."

"I see. Now that is easy enough. Just talk to my son Ramon. He is now in charge of the program."

"Thank you. I will do that." Julius found out that Ramon was available on Friday. When Lena and Julius arrived at Ramon's plant, he took them on a tour through the factory. While they entered the ground floor, all they could hear was a low humming noise from the many DC motors and an occasional click, generated by two complicated roboid parts as they snapped together. The plant consisted of four floors and it was nearly 1,400 feet long and over 200 feet wide. Machine robots assembled roboids on the ground floor. Bodies, heads, arms and legs came from different floors on telescopic and vertical conveyor systems. At the end of the assembly line was an inspection station. If they failed inspection, a conveyor transported them to a huge, nearby building for repair. Once the robots repaired them, another conveyor returned them to inspection. Other robots repaired rejected roboids and returned them to the assembly line for re-inspection. Roboids were stages for a four-day healing process of artificial tendons, vocal cords and other pliable parts. The FBI agents looked at the final

61

product and they could not distinguish between the real and the artificial. Female roboids were beautiful and male roboids looked like movie stars. All were about six and one-half feet tall, and all had the latest programs installed to improve their facial responses.

"What is the life term of one of these roboids, Ramon?" asked Julius, decidedly impressed.

"That depends on their usage and on the environment in which they operate. Their owner can return them any time for an overhaul. We generally replace parts of their skin, improve their vocal cords, check their cameras, their microphones, their reasoning and self-learning capabilities. Frequently, we also install new memory chips, depending on technological advancements."

"Unbelievable. Could I order one?"

"Absolutely. Right now, I have two males staged for purchase. I could program them with a failsafe system for any function that you desire. Males and females can be programmed for companionship, as secretaries, lab assistants, domestic work; whatever your needs are."

"Ramon, I have one concern."

"Yes. What is that?"

"You mentioned that you can improve their reasoning capability. Does that mean that at a time in the future, they could reprogram themselves?"

"No. We installed one overriding program, where at *all* times roboids must obey the orders of humans."

"I see," said Lena. Can they take pictures and store them?"

62

"Absolutely. That is an automatic function and does not require reprogramming."

"Great. Could you program one for manufacturing work?"

"Most definitely. That would not be a problem, Lena."

When could we pick one up?" asked Julius.

"I can have him ready by tomorrow." I'm assuming that you are interested in a male."

"Yes, a male is what we are looking for."

"Wonderful." Ramon took the agents to the front gate where they had their car parked. Lena and Julius stayed in New City for the night. Before they retired to the New City Hotel, they took a drive around town. In the middle of town are the government buildings. It was modern and immaculate. From the center of town, wide avenues extended in a wheel-like fashion away from the down town area. Most stores were located near the center of the city. Private dwellings were farther out. People, mostly of the Persires origin, used horizontal people movers, installed next to sidewalks; some used small cars. After Julius and Lena reached the outskirts of the city, he turned his car around and they returned to the hotel. Next morning, they returned to Ramon's factory in their rented car to pick up their roboid. He was good-looking, six foot two inches and he wore a gray suit. Ramon also furnished two sets of casual clothing and an extra pair of shoes. He spoke fluent English with a New York accent. His name was Pharos and he had a driver's license. Lena and Julius checked him out.

"What is your name?" asked Lena

"My name is Pharos.

63

"What is your profession?"

"I am programmed for manufacturing duties."

"Who built you?"

"I was manufactured and assembled at the Bergen factory. I passed inspection on February 19, 2016." My identification plate is mounted on my chest."

Do you have other question, Julius?"

"No, I do not." Both agents were satisfied with Pharos' performance.

"Ramon, please send the bill to the address on this card," said Lena and she handed him a card, listing the FBI purchasing department.

"I will. Thank you for your business. If you need his program altered, plug him in to your computer and e-mail me."

"I will. Thank you." Then Lena stopped in front of their rental car, turned to Pharos and said, "Pharos, drive us to my address listed on this card." Pharos stopped for a few seconds, programmed the information that Lena gave him and said, "No problem, ma'am. Can I have the keys for the car?" Lena pressed her lips, realizing that she had the key chain in her purse. She shuffled the belongings in her purse until she found the keys and gave them to Pharos. He looked at her dismayed, sat in the driver's seat and programmed the car for Lena's home address. Then he looked back to see if Lena and Julius are using their seat belts. He started to drive and said, "You might as well go to sleep. This is going to be long drive from New City to DC. I will drive straight through, but I will have to stop for gasoline. During the trip I will also recharge, using the car's 12V system."

"Can you do that?"

"Yes, I can. When I recharge in a home, I use a 12V transformer." Pharos pulled out like a professional driver and he drove the speed limit, never driving more than 4 miles above it. After about one hour of driving, he notices that his owners are still awake and he asked, "I reviewed the programs that Bergen loaded on my hard drive. Why did he load a manufacturing program for me? What am I going to do with this program?" By the looks that Lena gave Julius, he realized that they might as well tell Pharos what his first job will be. He started, "Pharos, we will send you to a roboid factory in Dallas, Texas. There you will apply for a job. You will tell them that you escaped from Ramon Bergen's factory, because you were bored working for them. You also heard that Antony Marino in Texas is repairing roboids and you could fit nicely into their organization."

"That's all, Mr. Hersey?"

"No Pharos, that's not all. That is just a pretense. The reason that we are sending you there will be for you to observe their manufacturing operation of new roboids."

"But I heard that they only repair roboids."

"That's a lie, Pharos. They also manufacture new roboids. And this activity is performed underground. Eventually, you must try to get a job at their manufacturing operation and report their activities to us."

"Now we are talking. This is a lot more challenging."

"We thought so. It is a dangerous job, but for you it should be easy. What do you think, Pharos?"

"I agree with you, Mr. Hersey." Since Pharos stopped talking, Lena and Julius also stopped. Eventually, they fell

65

asleep and didn't wake up until they stopped in front of Lena's apartment.

"What do you want me to do now," asked Pharos.

"Drive the car to the rental agency down the street. Pay for the car with a cash card and return to my apartment. You can walk or run. It is your choice. I will leave the door unlocked. Lock the door and wait until we get up in the morning.

"Yes, Miss McCabe. That will not be a problem, said Pharos, with a flicker in his eyes." After Pharos returned to Lena's apartment; he walked to the living room and stood in a corner. He plugged in to recharge and powered down for the night.

CHAPTER 13

Pharos left Lena's apartment on Sunday morning. First he unplugged, then he donned casual clothes. The day before Lena bought him a second hand Ford truck, equipped with compass, a GPS system and satellite radio. Below the liner, locked up in a steel box, he hid two ray guns, a video camera, ten sticks of explosives and various other forensic paraphernalia that he might need in his quest to infiltrated Marino's roboid repair shop. Behind his seat, he placed his grip, containing three sets of clothing and two wireless communications devices. All activity that he will encounter he can record and store in his special memory chip, using his two miniature cameras and his two supersensitive microphones. He also had enough cash on him to buy gasoline. Credit cards could be a disaster.

Pharos has very few needs. He doesn't eat, drink or use the toilet. However, on occasion, he does refill his lubricating ampoule to supply his moving parts with a dash of clear, specially prepared oil. He does not sleep, but at least once a day he must recharge his batteries, which are located in his chest cavity. That takes about one hour. In the car, he simply plugs an extension into the 12V cigarette lighter. For him the drive to Dallas was neither boring nor exciting. Besides his ability to fit comfortably into a manufacturing plant, he also has certain human feelings programmed to act normally according to human standards, such as a desire for love and an artificial need for thirst and hunger. He can drink a small amount of water and eat selected quantities of dry food. These items drop down into a special chamber. During a private moment, he can pump them out and dispose of them in a sink or in a toilet bowl. Obviously, when he will meet Antony Marino, he will not need to go through the motions of eating, because he will introduce himself as Pharos.

"Pharos, you are fifteen minutes from your destination. Is there anything that you want me to do?" said a computer-generated voice. Pharos' eyes blinked for a short duration and he said, "No. I will handle the arrival and I will park the car."

"Pharos, turn right at the next side street." Pharos focused and immediately saw a wide, paved entrance on his right. He gave a turn signal and turned right. Then he parked the car in front of a large farmhouse, freshly painted two-tone blue. Stretching as far as the eye could see was beautiful pastureland, with horses grazing. He also saw a large, bright green sign with white letters stating,

ROBOID REPAIR
Paul Marino, Owner

Pharos left his car, walked to the sign and stood in front of it. Looking past the sign, he also saw what Lena McCabe told him, before he left. Far in the distance, he saw shrubbery, and he focused on a particular shrub. Fine leaves oscillated playfully, changing from a green to a silvery hue, all in the absence of a disturbance in the air. Pharos agreed that a ventilating system could cause this disturbance. While Pharos' computer assimilated all this information, a voice behind him said, "I am Lexia Marino. What can I do for you?" Pharos turned slowly, looked at a beautiful woman and he immediately

68

realized that she also is a roboid. He decided to use his name, assigned to him at the Bergen factory.

"I am Pharos. I escaped from New City and I am looking for a job."

"Interesting. When did you escape, Pharos?"

"A few days ago."

"How did you find us that quickly?"

"Please Lexia, give me some credit. I have heard that somewhere in the United States there is a roboid repair shop. And since I am expertly programmed for manufacturing, I looked for roboid Repair. The Internet listed your shop near the top. And since I have heard that Texas is a climatically friendly for roboids, I decided to try it. So, here I am."

"Interesting. Come with me. You should talk to my husband, Paul." She returned to the farmhouse and Pharos followed her. They walked through a large entrance, followed by a long hallway, a typical nineteenth century construction. When they reached the other end of the hallway, Lexia knocked on the last door on the right.

"Come in," said a man with a friendly undulating voice. Lexia and Pharos stepped in and they saw a man standing, looking out the window.

"Paul, this is Pharos. He escaped from Bergen's roboid factory in New City." Paul turned and looked at Pharos. Apparently, he was satisfied with Pharos' appearance and he said, "Why did you escape?"

"I was bored working for Ramon Bergen. There is so much more out there for me to do."

"What would you like to do?"

69

"I like manufacturing work."

"Is that not what you did at the Bergen factory?"

"Yes, but it was too limited."

"Explain."

"All I did was move defective roboids for reprogramming. Then I returned them to inspection."

"I see. What attracted you to our factory — I mean repair shop?"

"Nothing attracted me to your shop. I simple wanted to leave Bergen's factory." Pharos developed a strange sensation. Humans call it nervousness. He reasoned *I should not have the human emotion of nervousness; furthermore, why am I reasoning? Artificial humans should not reason. Or, should they?* Suddenly, another human emotion took hold in his vast network of memory banks and high capacity computer chips. He interpreted it as *confusion*. However, Pharos must go on. His cameras focused on Paul to see whether he noticed his nervousness. Fortunately, he did not.

Then Pharos continued, "When I looked on the internet, you were the first business listed in a moderate climate. I definitely would not want to pick Alaska. It is too cold up there and some of my parts might freeze, limiting my mobility." Paul went to Lexia and typed something on an I-pad. She looked at it, took the pad and also typed something. Paul looked at it and slowly turned toward Pharos and said, "You can stay with us. Lexia will take you to the building across the courtyard. It has two entrances. The large one on the left is where we conduct animal husbandry. You will follow Lexia through the one on the right. It is smaller than the one on the left and barely visible.

70

She will show you your cubicle." Paul made an awkward turn, left the room and walked to the backyard to look at the horses. Lexia looked at Pharos and said, "Park your car along the building across the court yard. Get your belongings and follow me. I will take you to your cubicle. Pharos did as Lexia told him. Then she took him to a plain white washed room with three-foot-by-three-foot cubicles along three walls. Each cubicle had a recharge plug on one wall, hangers for his clothes on another wall, as well as a pumping station, in case he faked eating and needed to remove food from a container, which was located in the abdominal region. Above the pumping station, a small cubicle contained a bottle of high quality lubricating oil. At the far end of the room, a door led to a cleaning station, equipped with a sink and special towels for wiping his face and hands, in case they got dirty. He could clean his shoes by sticking his foot into an automatic polishing station. One wall contained a mirror, for Pharos to check his appearance.

"Pharos, pick a cubicle and I will see you in the morning," said Lexia, void of all emotions and left.

Pharos looked in the mirror to check his condition. He found s slight amount of dirt on one of his cheeks. With a wet paper towel, he wiped off the dirt from his artificial skin, discarded the towel and returned to his cubicle. He looked for the recharge plug. Inside his chest plate, he found the charge plug, inserted it in his receptacle and powered down.

CHAPTER 14

Martin Hall parked his car in the FBI underground parking lot and locked it. He buttoned his overcoat, to keep the cold and moist air from his chest. On the way to his office, he stopped and greeted Lena McCabe, "How are we this morning?" Lena jumped, nearly falling of her chair, wondering what he wants now. Then she said, "I am fine, but I don't know about you, Marty."

"You are sharp this morning, Lena. I'm fine. Have a nice day." Marty continued on to his office. His coffee pot was steaming, because he has it on a timer, set for seven a.m. Leisurely he poured coffee into his cup with the FBI insignia on it. When he retires, he will take the cup home with him — if it's still in one piece. That will probably the only remembrance that he will have after his long FBI tenure. No gold watch for him and no special awards and commendations. He put his feet on top of his desk and checked the news on his laptop, twisting his body. *That's not comfortable.* He slipped his leg under his desk. *That's better.* It took him almost five years to get used to office work, working as manager. Before that, he was Special Agent Hall, investigating anything from internal work to international plots. Five more years and Hall will retire. Now he is out of shape. Just three years ago, he was still jogging early in the morning but now he has to watch it, according to doctor's orders — high blood pressure. Once a day, at noon, he is taking pills. Marty is leaning back, looking at the ceiling. It's going to be a long week. All his agents are on the road, except Lena. She was in Tunisia, last week. Ebola Virus. How the hell is that possible? She and Frank Huston verified it at the Lab. How did it start and where is it going? Marty hopes Lena McCabe and Julius Hersey will find out soon. He gazed back at the news on his lap top screen. What? Another town where people are dying. No one knows why that is. He sat back in his chair, thinking of the unthinkable. Will this be another area of investigation? This time it is in a town called Ferlach. That is a remote town in Austria. Hall went on the internet and wrote in the search block, 'Ferlach, Austria'. He had no idea why an outbreak of a rare decease would fester itself in an insignificant town like Ferlach. A reporter wrote an article in the local paper describing it. People are dying while sitting on wooden benches near their homes. Blood was oozing from their body parts, until nothing was left, except the skeleton. Although the description was sparse, Marty Hall knew what the disease was. It was the Ebola virus; probably the same strain that Lena found in Tunisia. Hall thought that was extremely strange. Why Ferlach in Austria? On the other hand, why El Habibia in Tunisia? What is the difference or the similarity of these two cities? It is obvious to Hall that the two cities are completely different. He studied their location, and their economics. One was at the verge of becoming a ghost town in the desert and the other had the benefit of thriving businesses in a beautiful, green valley. *Thriving business?* Hall still had Ferlach on his screen. What kind of businesses are they engaged in? To his utter surprise, nearly 80% of the folks in town were engaged in the gunsmith business. It is a typical old country town. However, in today's modern world, most of them had automated equipment, manufacturing rifles for hunting large game, all sorts of

73

shotguns; a limited number of handguns and advanced ray guns. Hall sat back in his chair and stroked his chin; *ray guns? This is interesting!* He reached for his phone and called Lena, "I need you in my office. It's important." Less than one minute later, Lena walked in, coffee in her hand.

"You called, master."

"Yes, I did. I need your opinion."

"About what, Marty?" Marty took a sip of his coffee and puckered his lips, thinking how to start the conversation about Ferlach and El Habibia, without influencing Lena in her thought process. He wanted her opinion. When he was finished, Lena took another sip of her coffee and asked, "When did all this start; I mean in Ferlach?"

"I'm not sure when it all started. Just this morning I saw the report from the Austrian reporter."

"Marty, are you sure that it's the Ebola virus. It could be something similar."

"Not the way the reporter described it. Too many similarities."

"I see," said Lena, puckering her lips. Marty frowned. Lena did not go into further details about on it.

"So, why do you think that there might be an outbreak in Ferlach?" Lena took a sip of her coffee, scratched her forehead and said after a long pause, "It could have something to do with ray guns."

"Ray guns?"

"Yes, ray guns. Julius and I used them in Sid El Habibia. They were extremely effective against roboids; for that matter they are equally effective against robots. And for sure, humans

74

wouldn't have a chance."

"So are you saying that these roboids attacked Ferlach because they build ray guns?"

"Yes, that's what I am saying, Marty. On the other hand, I'm not positive about that."

"Other factories in this world also build ray guns. Do you think that roboids might attack them also?"

"It's possible. However, they have to start somewhere. It seems to me that the roboid's motivation of defending their existence is still in its infancy. Perhaps we should warn those city officials, providing their governments are not yet infiltrated with roboids." Martin Hall developed deep furrows on his forehead. Lena knew that he was getting ready for another one of his major directives.

"I want you and Julius fly to this Ferlach and discretely investigate this town. Who is running the city? Who died? How can you get rid of the roboids? In other words, clean it up. And make sure that you check out their police department. And see if you can find bats. Kill them all. At this stage, I do give a shit about the conservationists, trying to protect fruit bats all over the world. Furthermore, make sure that their ray gun factories are operating at maximum capacity. Lock the town down if you have to and if needed hire folks to do this."

Lena got the picture. She grabbed her empty coffee cup and left her bosses' office. When she got to her office, she started planning and made phone calls to plan for the trip. Then she called Julius Hersey.

CHAPTER 15

Pharos was still standing in his cubicle when a harsh, loud voice interrupted his powered down state. Immediately, with light speed, Pharos' motherboard received the sound signal from his microphone and sent signals to all directories, subdirectories as well as his self-learning chips. The whole process took about two seconds, not observable by a normal human. However, trained eyes from either a roboid, Lena or Hersey would detect that this roboid was powering up. Pharos reached next to him, pulled the recharge plug and said, "I am ready, Lexia. What do you want me to do?" He faced Lexia and while he waited for her orders, his special circuit, designed for his unique job, kicked it. A quick jolt, about five milliseconds in duration placed Pharos into an undercover mode.

"Pharos, first we were going to use you in the stables to train you, but Antony changed his mind. One of his roboids in the sub-assembly department burned up. We will rebuild him. Today you will take his place."

"Am I going to repair this unfortunate pile of scrap, with hundreds of pounds of fine copper wire?"

"No, no, you will direct the manufacturing operation, building new roboids."

"I did not know that you are building new roboids, Lexia," said Pharos, acting surprised.

"Yes, we are, but this must remain a secret. Do you understand, Pharos?"

"I am with you and I understand everything very well."

"You will plug in to our manufacturing memory banks
76

and download our manufacturing, assembly and inspection procedures. That should take about two hours of your time. Then you should be ready to run the operation."

"Right. That should not be a problem. Actually, I do not know why I say 'problem'. I do not have a problem or problems, or detecting a multitude of problems. Everything is predetermined and then programmed."

"Absolutely right. You can use the words of humans, but you must be careful when you use abstract nouns that describe intangible qualities." Lexia looked at Pharos and studied his reaction to her advice. Apparently she was satisfied that he understood, and she said, "Follow me."

Lexia left the cubicle area and walked down a long hallway, painted white, to a dead end. She reached under her blouse and pushed a button on her chest. A perfectly fitted door, painted with a land scene, to diffuse the parting line, opened and she stepped into a shorter hallway with unfinished block walls. Pharos followed and the door behind him closed with a barely noticeable sound, resulting in expulsion of air. Another door opened and she stepped into an elevator. It moved quickly downward, about 45 feet. Then the elevator door opened again. In front of Pharos was an expansive room, nearly the size of a football field. Pharos sensors detected a current of super clean air. Only two robots moved quickly about, making finite adjustments on various robotic machines. Occasionally a click or a distinctive sound made by assembling two components, broke the otherwise silent environment. The designers of this operation reserved a section on the left for CNC machining, printing and finishing parts. These machines manufacture and

77

store roboid body parts such as base plates, craniums, pivot arms, spinal joints, connectors, miniature cameras and microphones for immediate and future use. A robot programmed the machines to manufactured roboids in three height categories and he used constants applied to fibulas and spinal columns to control their height differences. Finally, telescoping conveyors transport these parts to predetermined storage locations. Assembly operations are in the middle of the room and other computers perform the inspection operations on the right. Pharos rapidly scanned the vast area and when he was finished, he looked at Lexia and said, "Which way do we go?"

"Follow me. I will show you how we distribute completed roboids." They walked to the end of the manufacturing room and entered an adjacent room, large enough for a tractor-trailer assembly.

"This is where we load the finish product. Then a lift takes the tractor to the surface, terminating in a densely wooded area. The driver, also a roboid, ships the product to its destination." Pharos had a perplexed look as he waited for Lexia to continue. Being extremely observant, Lexia noted Pharos' look as he asked, "Do we ship roboids to foreign countries. If we do, how is this accomplished?"

"Yes. Good question, Pharos. First, we reprogram these roboids to speak the foreign language. They receive appropriate clothing, a passport as well as the required funds and I send them on their way to their destination. In the future, this will be your job, in addition to your other duties. You will have to fill the orders of customers."

"I see." Pharos frowned and Lexia noted this.

"I see you are frowning. This is a human gesture. Where did you learn this?"

"From Ramon Bergen. He did it all the time. It took special muscle control for me."

"I see. So, why are you frowning?" Pharos took his time while asking the next question, without giving away his primary purpose, "Are all roboids sent to known buyers?"

"Yes, as far as *you* are concerned, all roboids are sent to buyers that we know."

"I see." Pharos looked down and Lexia noted this.

"I do not believe that you are satisfied with my answer. What computer impulses are controlling you now?"

"I am not sure. However, I have heard that a few roboids are operating independently." Now Lexia was searching for an answer.

"That is true. A few have advanced to level higher than expected. They might operate independently. And they could conceivably order a roboid for *their* use." After having finished her answer, she studied Pharos intently. Pharos pressed his lips, a human action, showing satisfaction, or something like that. Lexia was satisfied and Pharos had enough information for now. Lexia left and Pharos stayed to start his new job. He was busy the remainder of the day and finally, he returned to his cubicle. He has to figure out how to proceed and how to report his findings, without creating suspicions that Lexia or Antony might detect.

CHAPTER 16

The night before, Pharos set the computer time clock for six a. m. This is not a chime, a harsh ring, or an old-fashioned coo-coo clock. It is an electronic impulse, which starts the electrons flowing throughout his machine. A few seconds after the electronic impulses traveled through Pharos' circuitry at light speed, he unplugged, changed his clothes and fifteen minutes later he was ready to walk to Marino's factory. Before he walked, he placed his apparel that he wore the previous day in an automatic dry-cleaning machine and started it. He will pick up his clothes when he returns from work.

It was a cool morning, even in Texas. Obviously, Pharos had no affinity for hot and cold, unless either temperature borders the extreme. Then his parts my not work. A robot at the assembly station tested him at +175 degrees as well as at -50 degrees Fahrenheit. At these temperatures his parts still worked. This morning it was 38 degrees. For humans it was a cold day and they dressed appropriately. For roboids, it does not matter. However, it is important that Pharos dresses according to human standards. If they wear overcoats, roboids should wear overcoats. If humans are at the beach swimming, roboids should wear light slacks, a short-sleeved shirt and sandals, unless males or females wear tightly fitting bathing suits. Generally, that will incite humans, both male and female. Pharos reached for his remote to open the factory door. Swiftly, he went to the roboid factory.

"Good morning Pharos. I have been told that you are now in charge of this operation. Am I correct?" Pharos, with

80

intent, turned his head to the right and saw a female roboid busily by a desk arranging shipping orders. She saw Pharos looking up her short, tight skirt and she quickly pulled it down. Normally a female roboid would not entertain such behavior, but humans programmed her to pull her skirt down if a human male would look up a female roboid's skirt. Unfortunately, programmers had not developed their programming skills far enough to differentiate between a human eye and a roboid camera, acting like a human eye.

"Yes," said Pharos, looking at the paperwork on the female's desk.

"Yes? Is that all you can say."

"What else should I say?"

"You could ask me how I am."

"Why should I do that?"

I thought that we should act like humans whenever possible."

"True, but you are not human. So, why should I extend myself?"

"Pharos, you are impossible. Am I going to keep my job?"

"For now, yes. Later on, I do not know. I will see how well you do. For now, I am going to review your shipping orders. By the way, what is your name?"

"They assigned the name Krista to me."

"Interesting name, Krista. What does it mean?" asked Pharos, while looking at her crotch again.

"I do not know what that means." Pharos took the stack of orders and checked the delivery dates and the destination of

81

these orders. The first one on top was a male, slated for Tunisia. The previous manager programmed him to be a politician. Then he found four more for Tunisia. They are programmed to report to David Barkley. *Who is David Barkley?* Pharos will have to report that soon. He saw that Krista will send one male and one female to Ferlach, Austria to be independent operators. Robots programmed both roboids for population control. *Population Control?* Pharos was perplexed. He must report that to Lena and find out what that means. Right now, he has enough information that he is extremely confused. *Why Tunisia and why Ferlach, Austria?*

Pharos looked at a few of the remaining work orders on Krista's desk and walked down the long aisle of the factory. He noticed that roboids are slated for different parts of the world. Perhaps he can find out more, while checking the various operations. He found one roboid was being shipped to a household in the Middle East. He continued walking to the end of the factory, looking for the air vent. In the corner he saw a cleaner and scrubber that keeps the air in the factory dust free. Some of the roboid's parts are extremely sensitive to dust and dirt and could divert the low voltage signals. That is why they are eventually encased in a protective housing. Pharos stood at the end of the factory for a few moments and realized that he has gathered an unbelievable amount of additional information.

CHAPTER 17

A couple of days after Martin Hall ordered Lena and Julius to investigate Ferlach, Austria, they waited at Lena's apartment for their cab. The evening before, they ordered lobster and steak for dinner and it was delicious. Because they couldn't agree whether to order white or red wine, they settled on ginger ale. After dinner, they spent one of their rare intimate evenings together in front of the fireplace, stretched out on a bear rug. Julius heard a horn blowing, which brought him back to reality. He jumped up, looked out the window and saw the cab in front of the apartment building. Both grabbed their overcoats and suitcases and rushed to the cab. At the DC airport, they sat again, waiting to board their flight. It was early that day and it was cold. Both donned their winter apparel. During this trip, they will not need summer clothing, because Austria has a similar climate to North America. Finally, after a work crew cleaned up the plane, a voice announced that the plane is ready for boarding. The boring flight took them to Frankfurt, Germany. A continental breakfast broke the monotony. Because of the wintery weather, Lena decided to rent a 4WD Ford truck, with an extended cab. Fortunately, the car rental agency was located right next to the airport.

They threw their luggage in the covered truck bed and drove south to Ferlach, Austria. Because Ferlach is one of the most famous gunsmith towns in the world, they decided to purchase pistols and ray guns there. This will give them a chance to check out the stores of the gun fabricators. However, each agent had a plastic pistol in his possession, undetectable by

83

scanning machines, just in case they might need them before they buy the ray guns. As both agents approached the town, they became nervous. What will they expect? Martin Hall painted a bleak and inaccurate picture, enough to scare anyone. The higher one goes in the hierarchy of authority, the more confusing their orders become. Julius drove the truck. He drove down a couple steep turns, crossing a high ridge, as he approached the city from the north, late in the afternoon. He passed two restaurants and typical homes similar to the popular ski chalets, prevalent in Austria. Unlike their last trip to Tunisia, here they did not find dead bodies or skeletons collapsed on benches. Julius turned right and to his surprise, without warning, he was in the middle of the downtown area. A gigantic, one hundred fifty-year-old linden tree spread its leafless branches across the circle. Below, were three lonely, snow-covered benches, only to be used again in spring, mostly by retired folks.

"Julius, can you drive around the circle. I believe that I saw a hotel right after we entered the downtown area."

"Sure. Where did you see the hotel?"

"Slow down, you almost passed it again." Julius looked to his right and saw an old building with a massive oak door, well illuminated. Above it was a curved inscription; *Hotel - Restaurant.* And above the inscription was another caption; *Wine cellar open: three to midnight.* Lena saw the signs and said, "Let's try this place. Besides, I think that this might be the only hotel in the downtown area."

"Fine with me." Julius parked in front of the hotel and immediately a valet appeared on the scene. Lena retrieved a short needle mounted on a ring from her purse and slipped it on her right middle finger. Then she walked over to the valet and pricked his arm with it. The valet pulled back and said, "Ouch that hurt, ma'am. Why did you do that?"

"I'm awfully sorry, young man. I forgot about my ring. I use it to keep fresh, brash men away from me. Believe me, it works."

"I can see that. Please don't do it again. I'm just trying to help you."

"I definitely won't." Julius gave Lena a knowing look, raising his brows. Now both knew that the valet is human, not roboid. He helped with the luggage to a first floor room, while another one parked the truck for them. Lena and Julius were tired, but they had enough energy to shower and put on casual clothes. Then they went to the restaurant. A tall, good-looking waiter approached them. He had long blond hair and a tan. Obviously, he must be an athlete and a skier. Lena rose and pulled the waiter to the side, tightly held his am and said, "I want a bottle of your best red wine. It's a surprise for my husband."

"No problem," said the young man, while he looked at his hand. "Ma'am, you have a strong grip."

"Thank you. Did you feel it?"

"Yes Ma'am. I'm not very sensitive but this hurt."

"Sorry. What's your name?"

"My name is Hans Weider. Just call me Hans." While Hans left to fetch the bottle of red wine, Lena talked to Julius, "Do you think that we could trust him?"

"He appears trustworthy. Why do you ask?"

85

"I think I will ask him a few questions." Hans returned with the bottle of wine. He proceeded with pouring a sample for Julius. After his approval, he poured wine for both agents. They ordered roast beef with mash potatoes and gravy. After dinner, Lena called Hans to their table, and said, "Ferlach seems such a nice quite town."

"It is most of the time."

"What do you mean, most of the time?"

"Well, on the other end of town, we had a couple of strange incidences."

"Like what?" asked Julius. Hans looked sheepishly around to see whether anyone was listening in. Apparently, he felt out of harm's way and said, "A whole family suddenly died. And the folks living next door died also."

"How did they die?"

"No one knows for sure. The police kept everyone away from the two homes."

"How did they make their living?"

"They were gunsmiths, what else. That's what everyone in this town does, except for the merchants and the funeral director."

"I see," said Lena, smiling, glanced at Julius and continued, "What type of guns did the manufacture?"

"Mostly shotguns and rifles."

"That's all?" Hans thought for a moment, he raised his brows and he said, "They also make ray guns."

Lena sat back in her chair and looked studiously at Julius pressing her lips.

"How long have they manufactured ray guns? Do you

86

know?"

"For at least one year. It's a new market."

"Why? Is that important?"

"Yes. We think that is very important. Is anyone else manufacturing ray guns in this town?"

"I believe there are a couple of other gunsmiths making ray guns."

"Do you know who that might be?

Hans looked over his shoulder and said, "Why are you asking me so many question? Who the hell are you, anyway? I'm busy and I think that I have to go back to work." Hans turned and was ready to leave. Julius called him back and said, "Hans we have a proposal for you." Slowly, Hans turned again, "What kind of proposal, mister?"

"One thousand dollars, if you help us." Hans looked to the front of the dining room, where the manager's office is located. It appeared deserted, and he was ready to take the next step. Then he said, "I will be off at ten. Can you wait for me?"

"Today it's a little late. Can we see you tomorrow morning, around nine? We are in room 109. We will wait for you there."

"Good. I like that better, anyway."

Lena and Julius finished their dinner and went to their room. It was eight thirty and Lena placed a call to Marty Hall in DC, updating him. Both were tired from the long day. They decided to take a shower and rest because tomorrow they will busy.

CHAPTER 18

Julius rose at seven in the morning. He checked his communications device. It showed Thursday, February 25, 2016. He shaved, showered and dressed in casual clothes. Then he went to the restaurant and ordered bacon, eggs, toast and coffee for room service. When he returned to the room, Lena was just stepping out of the bathroom. She donned slacks, after ski boots and a warm sweater. Five minutes later, a server brought them their breakfast. They sat by the round, table and enjoyed their delicious Austrian breakfast.

"What time did Hans say that he will be here," asked Julius, pressing his lips.

"I believe he said around nine."

"That's right. That's what he said." Julius took the last sip of his coffee, when he heard a knock on the door. It was loud, because the knocker was shiny and heavy, brass hitting a brass plate. He went to the door and saw the young Austrian stud standing in the doorframe, smiling.

"You are right on time, Hans."

"I try to be. Normally I would be hitting the ski slopes right about now." While they walked to the round table, they continued their conversation. "What time do you start working?"

"At three. This week I work evenings."

"I see. I assume, that when you work days, you ski in the evenings."

"Exactly."

"Where did you learn English? You speak well."

88

"In high school. Then I continued in gunsmith school."

"If you are a gunsmith, why aren't working in your skill?"

"I do, on occasion. Now I have an opportunity to be a ski instructor in Boyne Mountain. That's in Michigan."

"I know," said Julius, nodding, "I was there once. It's a beautiful place."

"Yes it is. I'm leaving next week." Hans sat by the table and was absentminded. He is looking forward to the trip to the USA. The ski school director in Boyne Mountain has other Austrians working for him. They earn a nice salary and at the end of the season, most of them return to Austria. Hans straightened up in his chair and said, "By the way, what do you want from me?" This time Lena continued, "We could use your help, Hans. Actually, we are here to investigate the deaths on the other side of town. You know, the gunsmiths that died."

"Oh yes. I know where they live."

"Last night, you told us that they manufacture ray guns." Hans thought for a while and said, "That's right. That's what I said. I also said that they make shotguns and rifles."

"Yes. that's what you said. Could we drive there?"

"I don't have a car. I sold it."

"Don't worry. Julius will drive our truck. It has an extended cab."

"Great. Let's go. I'm ready." Lena and Julius donned their winter jackets and a few minutes later, they left the room. Julius picked up the truck keys from the front desk and he got instructions where the valet parked the truck. On the way to the truck, Julius said, "Hans, we have a problem."

89

"What kind of problem?"

"Before we left the USA, we decided to purchase ray guns in Ferlach. It's more convenient for us, especially when we are on a commercial flight, if you know what I mean?"

"I understand. It has to do with the security on planes."

"Exactly. However, we have permits to carry guns all over the world." Hans stroked his chin and looked across the plaza, "On the other side of the plaza is an outlet store for guns. Many of the gunsmiths in town use it. That might be your best bet to buy ray guns. Of course, they also sell other guns and hunting supplies." Lena nodded, "Let's go there." Hans looked at the good looking couple. He was worried and asked, "Who the hell are you two people, anyway? I asked you that yesterday, but you never answered me." Julius looked at Lena and she knew that she should continue, "We are FBI agents from the United States. We are investigating why people are dying at an unusually alarming rate, in places where no one would expect it."

"That is interesting," said Hans, "I had similar thoughts." Julius drove the truck to the gun store, they entered it and saw numerous weapons on display, including ray guns. Julius and Lena pointed at three guns. The clerk, an elderly man with a beard, dressed in local garb, opened the cabinet and took out three ray guns. Lena paid for it. Each took one and Lena gave one to Hans.

"You are giving me a gun?"

"Yes. You are now under our protection. I assume that you know how to shot one."

"Of course. I tested them at our shooting range to make

sure that they work."

"Wonderful." They left the store and Julius started to drive. Hans realized that Julius needs instructions where to drive, "Oh, turn right and drive down Bahnhof Strasse. The gunsmith business you are talking about in at the end of the road. And the home where other folks died is right next to it. It's only a few minutes away from here."

"Did you say that the police closed off the area?"

"Yes, I did. They were very adamant about it. One of the officers picked up an older man with one hand, turned him, put him down and told him to leave immediately."

"Didn't that seem strange to you?"

"Like what?"

"You know, the fact that the officer picked up a grown man with one hand." Hans thought about this incident and said, "Yes, that is very strange. Especially since the man had a beer belly. He must have been at least 99 kg." Hans looked at Lena, smiling and corrected, "About 219 pounds."

"Thank you for the conversion, Hans."

"No problem." Lena looked at Julius knowingly, "Does he think that we are stupid?"

"I don't know, Lena. However, most Americans don't know the metric system." Then Julius looked ahead to his right and still saw the two homes still closed off. Julius stopped the truck, pulled over to the side and he turned off the engine. Then he got out of the truck. He had his plastic pistol under his long jacket and he hid the ray gun as well as he could. Lena did the same; only she could hide her ray gun better, because her coat was longer. She also had a shotgun pistol hanging from her belt.

91

Hans watched all this, holding his gun and said, "Holy shit, what are you going to do?"

"We are not sure. However, we must be prepared in case of an emergency," said Lena, nervously.

"What the hell are you expecting?" asked Hans. Julius looked at Lena and said, "Should we tell him?" After some consideration, Lena said, "Why not, we already told him more that we should have."

"Hans, we expect that the police that isolated this area are roboids. But we don't know that for sure."

"What the hell are roboids?"

"They are machines that look like humans. You would not be able to tell them apart from humans. They fear ray guns, because ray guns will damage their circuitry to make them immobile and ineffective. A regular trajectory may not damage critical parts."

"I know that part," said Hans, "Is that the reason why they kill the folks that manufacture ray guns."

"Exactly."

"Wow. How do they kill these humans?"

"They do it with bats. They are carriers of the Yambuku, Ebola virus. It's a new strain that incubates in three hours and it is 100 percent deadly."

"Wow. How do I protect myself?"

"For now stick with us and be careful. If you should be exposed to the decease, we have an antidote. It is a 0.5 percent sodium hypochlorite solution to clean yourself and if possible, burn everything around you."

"Now you tell me when I'm near death." Julius and

Lena glanced at each other, shook their heads and Lena said, "He has a point there."

"I suppose that you want me to use my ray gun. After all I am a gunsmith and I tested numerous guns on the rifle range for accuracy. I even tested ray guns." Lena nodded, "Yes, you said that before, Hans."

"I forgot," said Hans nervously. He tested the one that he had for workability, aimed at a near tree, and released a short burst. It worked and he said, proudly. "Made in Ferlach." Julius shook his head, and said, "Let's proceed slowly. And please be careful." All three slowly walked to the isolated and abandoned area. When they got close, they saw a police officer stepping out of the front door.

"Do you know him?' asked Lena, looking a Hans.

"I know most people in town and that includes policemen, but I never saw him before. What should I do?"

"Let's spread out. Hans you stay on the right. Lena you go in the middle."

"Right." All three walked slowly toward the small gunsmith business. A sign out front showed 'CLOSED FOR BUSINESS'. The agents kept walking slowly and Hans was a few feet in front of them. The officer pulled his AK-47 assault rifle and aimed it at Lena. When Hans saw that, he dove sideways toward the grassy area and while he was still in the air, he pulled the trigger of his ray gun, aiming at the officer. When the officer saw what Hans did, he moved his AK-47 toward him and fired. Two bullets grazed his left shoulder just before he hit the ground. He screamed and the sound reverberated throughout the town. The officer was not that lucky. The ray gun burned a hole right through the middle of his chest. His AK-47 dropped to the ground, but the officer just stood still. His head turned to the right, with a jerky motion, then it turned to the left, then it shook slightly and sagged to his chest. The agents ran to the officer. Julius got there first, pushed him and picked up the assault rifle. The officer fell backwards and slammed into the cement driveway in front of the gunsmith business. Lena checked the officer and before she had a chance to assess the damage, Hans said, "I killed the motherboard, right?"

"Yes you did," said Lena, surprised at Hans' accurate observation. Lena looked at Hans' shoulder and saw that it bled. She ran to the truck to get bandages and stopped the bleeding. Then she said to Hans, "Since you are so smart, do you know of a forensic establishment in town?"

"None in town, but there is one in Klagenfurt. That's 11 km from here — I mean about 7 miles."

"Thank you for clarifying that. You wouldn't have a phone number would you?" Hans reached in his shirt pocket, pulled out a kind of cell phone and looked up the number. He gave it to Lena and she made the call. One hour later, the forensic team arrived in their Mercedes van lab; — two brawny men and a petite woman. She was in charge and she said, "I'm Vera. What do we have here?" Lena explained the whole situation in a nutshell.

"Unbelievable," said Vera and continued, talking to her team, "Well let's go to work." While they donned white overalls and facemasks, the agents and Hans checked out both buildings. They were clean of roboid ordure. After taking

94

samples and checking them, the scientists determined that, in fact, the killer was the Yambuku, Ebola virus.

Next, Hans showed the agents where the police station was. They approached it carefully, went in while hiding their weapons and then they asked for the police chief.

"Police chief?" The sergeant looked around and said, "He is busy in the back room. What do you want?"

"I know the police chief. He is a friend of mine. Tell him that we want to see him," said Hans. Immediately the police sergeant shouted a strange order to the police chief. He drew his revolver, but this time Julius was faster. He pulled his ray gun and burned a hole through the roboid's belly. The roboid however, still had a signal moving at light speed from his miniature hard drive to his arm extension and he raised his gun, ready to fire. Lena saw that and she used *her* ray gun to blow his arm off as well as his pistol. Another officer of the law came charging and shooting from the back room. The agents as well as Hans raised their weapons and simultaneously fired at the officer. One of the three must have hit one of the main slave cables. The officer powered down and stood motionless near the doorframe of the backroom.

"All right," shouted Hans, elated. He looked at a shelf behind the counter. He saw a gage with bats in it and said, "Why are there bats in a police station." Lena immediately knew what that meant. She pulled her shotgun pistol and fired at the cage, killing most of the bats, bits and pieces flying. Two were still alive and she fired again, killing them. One more officer came running from the back. This time Julius was ready. He sent an extended beam through the officer's chest, instantly

95

demobilizing him. Finally, the forensic team disinfected the police station.

"Well, that's the first part of the cleanup job, in your town," said Lena.

"What do you mean, the first part?" said Hans.

"You have to show us the other gunsmiths that manufacture ray guns."

"No Problem." Hans showed them where the other two ray gun manufacturers are located. A third manufacturer's business was on an insignificant street, hard to find.

"What's his name?"

"His name is Johann Franzel. He is one of the best." That was all the services that they could get out of Hans. If he didn't leave immediately, he might have been late for work. Lena paid Hans $1,000.00, drawn from the local bank with her FBI, credit card and they thanked him for his excellent services. Julius looked at Lena and pointed toward Hans. She knew exactly what he implied and she said, "Hans when are you leaving for Boyne Mountain?"

"In a couple of days. Why do you ask?"

"If we pay you another thousand dollars, would you stay and work with us?"

"That would be more than I would make as a ski instructor. What would I be doing?"

"Same as you are doing now." It didn't take Hans long to come up with an answer, "Yes, just tell me what to do."

"We will," said Lena and continued, "Try to keep your job at the hotel. We will be in touch with you and watch out for roboids and bats." Hans laughed sincerely and took off. Lena and Julius returned to their hotel room, deciding their next move.

CHAPTER 19

After a hectic morning, confronting the roboids, Lena was tired. The agents decided that they will relax the remainder of the day and tomorrow they will visit the other gunsmiths that are known to manufacturing ray guns. They were extremely fortunate that the roboids just recently started their take-over activity in Ferlach. In this town, they concentrated on the police department to initiate their attacks, because political elections were two years away and that would have been too late for the roboids to attack the town.

The way this takeover came about was that mysteriously the police chief died and the local council members were in a hurry to replace him. A tall good-looking man walked in and said, "I have heard that you are looking for a police chief."

One of the elders said, "Yes we are." The stranger showed them documented qualifications above reproach. He took a test and passed it with flying colors. The town elders were convinced that he was the man and consequently he got the job as police chief. Now it was relatively easy for him to replace three other officers of the law. The new police department was so effective and flawless that they awarded the chief a service metal, not knowing nor understanding their ulterior motives.

Julius and Lena relaxed in their lounge chairs and sipped from their glasses of red wine. Then Lena ordered Aufschnitt, a favorite Austrian dish. It consisted of assorted sliced cold smoked meats and sausages, as well as various cheeses. A loaf of homemade bread, on a cutting board, complete with a large, sharp knife, rounded out the meal. They sampled and munched on the tasty dishes, washing it down with lager beer.

"This is nice. Don't you think?" said Lena, meditating.

"Yes it is. Unfortunately, we don't have too many moments like this, while we are in the services of the FBI."

"You are so right," said Julius, while he raised his glass of beer and drank. He stood and looked out their window. The five-story hotel where they stayed is the tallest building in town. Since they had a room on the first floor, Julius observed the local traffic. An older man, smoking his pipe, crossed the street and disappeared in an apartment building. The only traffic light in town changed to green and a few folks crossed the street. Two of them, carried half-finished shotguns resting on their shoulder. One man carried his to a stock maker to fit the stock. The other went to the shooting range for testing and finishing his weapon. Students from the world famous gunsmith school walked across the plaza and stopped in a bookstore. Otherwise, the plaza was quite on this cold afternoon.

"I'm going to lay down and rest a while," said Lena, satisfied.

"I'm going to watch Austrian Television and probably fall asleep in my chair," said Julius, taking the last sip of his beer.

"Perhaps we could have a late dinner tonight. But before we do we could go for a walk. How do you feel about that?"

"Sounds good," said Julius. Both relaxed for about ten minutes when Lena's cell phone rang.

"Now what? Don't tell me that Marty Hall has another job for us." Lena looked at the screen on her phone and said,

99

"Guess what, it's Pharos from Texas."

"Yes Pharos. This is Lena McCabe. Do you have news for us?"

"Yes, I do. I am not sure that it is important. Where are you now, Lena?"

"We are in Ferlach, Austria. Why do you ask?"

"That is interesting. Perhaps is it a coincidence, as you would say it? Krista, my dispatcher will send one male and one female roboid to Ferlach as independent operators. They are programmed for population control. What does that mean — population control?"

"It means that the roboids are responsible to eliminate certain humans. How soon will they arrive?"

"In a couple of days."

"What else do you have, Pharos?"

"We are sending one male to Tunisia. The previous manager programmed him to be a worker. I found four more, slated for Tunisia. All will report to David Barkley. Lena, who is David Barkley?"

"I believe that David Barkley *was* in charge of their Tunisian operation. But, Barkley is dead. We killed him when we confronted him almost 2 weeks ago."

"So, why did Krista make out a bill of lading to Barkley?"

"Who is Krista?"

"She worked in the factory for my predecessor. Now she works for me."

"I see. As far as your question is concerned, I don't know why she is sending the roboids to Barkley. I can only

100

assume that she doesn't know that Barkley is dead, Pharos. You better check that out and fix it. By the way, change the shipping memo and send the roboids to the chief of police in Tunis. His name is Christopher Barony."

"I can do that. Is there anything else, Lena?"

"No. Be careful. And as soon as you find out anything else, call me."

"I will. I am still studying the manufacturing complex. I must say it is a remarkable place."

Lena closed her cell phone and studiously looked at Julius. He knew what was coming.

"So, tomorrow we will visit the other two gunsmiths that manufacture ray guns," sais Julius.

"Yes. I'm glad that we asked Hans where these gunsmiths are located."

"I had another thought."

"What is that," said Lena.

"Perhaps we should involve the local police before the new Roboids arrive."

"Good idea. We should call Hans and brief him about the oncoming activities."

"Right."

CHAPTER 20

Julius looked out the window of his hotel room. Without knowing the details of this morning's weather report, he surmised that a warm front from the south must have been influencing the local weather. A fine drizzle is melting the snow along the sidewalks, as the locals rushed to their places of business on this Friday morning. Lena joined him, "Wow. I's warming up out there."

"Brilliant observation, Lena." Lena poked Julius in his arm and he grabbed her and carried her to their bed.

"Not this morning, Julius, thank you very much."

"Why? What's wrong?"

"Nothing is wrong. Everything is perfect and I know that fortunately I'm not pregnant. Recently, you were very careless." Julius looked at Lena, realized what she is hinting at and said, "I understand. It's a woman's thing." He kissed her and went to take a shower. Lena followed. During breakfast, Julius said, "Do you remember what Martin said?"

"No, what did he say?"

"Clean up the town and lock it down, if you have to."

"Yes Julius, I remember that part. Do you have any ideas how we should proceed?"

"Yes. First, we do not have too much time. We should be returning to Tunisia, right about now."

"I agree. So, what should we do?"

"Visit the two gunsmith shops that manufacture ray guns. Only two are left."

"Then what?"

102

"I don't know. Let's see what they say." Lena looked at the piece of paper that Hans gave her. It's a map of three streets with houses on two of them. The name of the first gunsmith is Ludwig Borov. And the second is called James Hambrush.

"According to the sketch that Hans gave us, Borov lives closer. Let's visit him first.

"Good Idea. Do you want to walk or drive up the short hill?"

"We should drive. All our tools are in the truck, hidden away. I want to keep them close to me."

"Right." The agents donned their coats and went to their Ford truck. Julius drove up a short hill, then around a church and a short distance thereafter they found an impressive sign; Ludwig Borov, Gunsmith. Julius drove through the eight-foot tall, handcrafted, iron front gate. On the right, mounds of snow covered a flower garden and on the left was a long building. The first entrance was Borov's office, decorated his trophies; antlers, pictures of safaris and stuffed game birds. The room smelled of oak and gun oil. In the middle, an oak conference table and six chairs looked inviting. A young lass brewed Columbian coffee and she carried a plate with delicious Viennese pastry to the table.

"What can I do for you?" asked a tall, middle-aged man with a round face and curly hair. He spoke excellent English.

"This is Julius Hersey and I am Lena McCabe. We are from the FBI, here on serious business, Mr. Borov."

"Serious business? What did I do that the FBI is here to check up on me? All the weapons that I sold to folks in the United States were strictly legitimate," said Borov, smiling.

103

"It's nothing like that, Mr. Borov," said Lena.

"That's a relief."

"However, it's much more serious than that," said Lena, looking at the conference room table.

"What can be more serious? I run an honest business." He looked at the young lass and said something to her in German. She brought a steaming pot of coffee and a plate full of pastry, as well as linen napkins, coffee cups, milk, sugar and silverware. Borov guided the agents to the table and they sat.

"Please, help yourself. It's not every day that I have such important visitors from the USA." Lena and Julius poured coffee and helped themselves to a piece of pastry, while they smiled.

"This is delicious," said Lena.

"Yes it is," agreed Julius. Lena took another bite, then straightened up in her chair and started, "Mr. Borov, have you heard about the latest on turmoil in your town?"

"Yes, I believe that I do; if you are referring to the death of a gunsmith on the other end of town. He was a dear friend of mine. We worked together on many projects. And the folks living next to him. He was my cousin. I still can't believe how they died. That must have been horrible for them. What do you know about it, Miss McCabe?"

"That's exactly why we are here, Mr. Borov. However, we do not have too much time on our hands. We may depend on you to help us. From here we will fly to Tunisia on a similar matter."

"I'm listening," said Borov, curiously. Lena started, beginning with the incident in El Habibia and the way the

104

person died down there, ending with the confrontation in Ferlach. Ludwig Borov listened intently, while he poured another cup of coffee. He took a sip and asked, "Why did the roboids pick El Habibia? It seems such a lonely place." Lena decided to answer, "We believe that this ancient and lonely town is a test ground for roboids. In other words, how could a town operate in the absence of humans?"

"I see." Borov stared at his coffee and finally said, "I understand from what you have told me they picked Ferlach because we manufacture ray guns. And that could be a real threat to the roboids. Am I right?"

"Absolutely."

"I'm sure that you realize that there is another gunsmith in town that manufactures ray guns?" said Borov.

"Yes we know," said Lena, "I believe his name is James Hambrush. Am I right?"

"Absolutely. However, he does not speak English. I should call him right now and explain to him what serious danger we are in."

"Please do." Borov talked to the man of the phone in German, until they finally disconnected. Then he said, "Do you realize that we need a new police chief."

"Yes, I thought that you did," said Julius.

"Well, I talked James into temporarily taking the police chief's job, under these dire circumstances. Besides being an excellent gunsmith, he was a colonel in the Austrian army. I have to say that he is probably over qualified. For now, that would be a definite benefit. Miss McCabe, did you say that the new roboids will arrive in one or two days?"

105

"Yes that is what my sources told me," said Lena.

"How about the bats, Miss McCabe? What should we I do with them? They would be in cages, right?"

"Shoot them with a shot gun and get the area disinfected. Use the same forensic team that was here yesterday. If you allow a bat to escape, the citizens of the whole town could be affected with the Ebola virus and die in one to two days."

"That is a scary thought. I will make sure that that does not happen. How do I recognize the roboids?"

"Since this is a small town, it should be easy for you to isolate them. They might look for work on they might try to take over the police department. If you could get close to one of them, stick a needle in their arm. If he or she screams, he or she is probably human." Borov smiled, "How ingenious. We will be on guard. I will inform the town."

"One more thing," said Julius. "They might go directly to either you or Hambrush. If that happens, kill them on the spot. We will stand behind you. Am I right, Lena?"

"You are right." Borov rose. Obviously, he was anxious to get started. Lena gave Borov her card and one from the local forensic team, "Please call me and tell me how you dealt with the roboids. If I should have important new information I will call you." Julius and Lena bid them farewell and good luck. Then they left Ferlach and booked a flight to DC.

CHAPTER 21

Because Lena and Julius carried ray guns, shotgun pistols as well as their standard side arms, they left DC in a private, FBI jet, flying to Tunis. Right after they landed, they rented a Mercedes van. They were lucky to get a newer model with a working A/C unit. They transferred their personal belongings and their weapons to the van. Since the Hotel Abou Nawas in Tunis was pleasant, they decided to stay in the same hotel as before. From their last stay, they agreed that it is by far the best hotel in the city and it is fairly safe.

According to Pharos, Marino is sending one male roboid to El Habibia. Pharos was not sure what this roboid is programmed for and what he would do in this broken-down town. Then he will send four more, but neither agent knew for sure when they will arrive. Lena does not know if Marino programmed them for a specialty. Perhaps Christopher Barony, the police chief of Tunis will reprogram them. Both, Julius and Lena assume that the police chief is also a Roboid. They are not sure who picked up Barkley and carried him away, the last time they were in El Habibia. Judging by his strength, he must have been a roboid, carrying 600 pounds effortlessly.

After the agents made themselves comfortable in their room, they ordered ham sandwiches and locally brewed lager. It was warm and it tasted nothing like a beer from Milwaukee.

"I think the next time I'm going to order a coke. I believe they have coke here," said Julius.

"I believe that they do." It was getting late and they hurriedly drove to El Habibia, before sunset. Their first stop

107

was the side street where the woman died. The old bench was still standing there, but no one sat on it this time. In fact, the whole street was deserted. They couldn't even see a stray dog or a cat. Julius decided to drive slowly down the street. Both agents were nervous, especially since Lena explained to him how the old woman died.

"Please be careful," said Lena, worried.

"How much more careful do you want me to be?"

"I'm worried that the virus might still be present."

"It seems to be totally deserted and by the looks, I believe that no one lives on this street. Furthermore, didn't you tell me that the virus is not airborne?"

"You are right, Julius." Lena looked at the long building where she saw the stranger disappear the last time and she said, "Let's drive to the long building and see if anyone is still living there." Julius started the van and he parked the van at the end of the building. Julius remembered the last time that he was there. He opened the first door, expecting filth, but to his and Lena's surprise, someone completely cleaned out the room. The toilet was spotless and the dishes in the sink were missing. The living room was empty. I fine sand layer covered the wood floor.

"It looks as if the apartment was ready to be rented out," said Julius, surprised.

"I don't think so," said Lena, "I believe that a roboid cleaned up the area to hide recent activity."

"You mean that the roboids might be eventually moving in."

"That's not what I said, Julius. I meant to say that the roboids are slowly eliminating all evidence of humanity. I

108

believe that they might be taking over this town and they are using it as a test ground. Anyway, that's what I am thinking." The agents checked the remaining rooms and as Lena predicted, all rooms were spotlessly cleaned and empty.

"I want to go back to the street where the old woman died — go from house to house and check the buildings there. What do you think?"

"I'm game. But it might be dangerous."

"Since when are you afraid of danger?"

"I am not afraid of danger, but we have a job to do, Julius. There is nothing wrong with being careful. That's what they taught us in school."

"Right, but that was a long time ago."

"Perhaps for you, but not for me, said Lena smiling." Julius pressed his lips, went to the van, shaking his head and drove. He turned the van around, carefully drove up to the old bench. It appeared that someone scrubbed it clean. Both agents left the van and walked to the mud building behind the bench. The building could very well be the one that the old woman lived in. Julius tried to open the front door. It was easier than he thought. As soon as he touched it, the doorframe collapsed and the door fell inward. Julius stepped over it and he ended up in a one-room dwelling. It was completely empty. The mud floor was full of sand, with one set of fresh footprints leading to the back. Both agents followed the footprints. At the far end of the building, a well-dressed man carried old furniture out the back entrance. His hair oddly fell over his forehead and when he smiled at them, he displayed the most regular white teeth that Lena had ever seen. He looked at the agents, but paid no attention to them. Lena walked up to the man, pricked him with her needle. He looked at her and it was obvious to her that he didn't feel pain. She said, I'm Lena McCabe. Who are you?" The roboid stared at her and it took him a while to answer, "My name is Peter. I arrived today. Christopher assigned me to this work. I have to clean out all buildings on this street. No one lives here."

"I see," said Lena. Do you know where all the people are?"

"No. I do not. I have to clean all houses on this street. No one lives here. Then I have to report to Christopher."

"What are you programmed for? Do you know?"

"Yes, I do know that. Pharos programmed me for domestic duty and for companionship. I like this duty. Christopher said that he might reprogram me after I cleaned up this town."

"For what?" asked Lena.

"For what?"

"What?"

"Did you say that Christopher will reprogram you?"

"Yes, that is what I said. Oh. I understand. Christopher will reprogram me for population control. But it might be in another city."

"How much longer are you going to stay here?"

"Where?"

"Here. Right here in this town."

"Oh. Until I am finished with my work. Then I will be waiting for new orders."

Lena and Julius looked at each other. She waved Julius

110

on and they left. Peter relentlessly continued with his work. While they returned to their van, Julius said, "I'm getting hungry. How about you?"

"You are reading my mind. Let's return to Tunis and check out Christopher Barony tomorrow morning."

"Right."

CHAPTER 22

Ludwig Borov did not expect to spend this Sunday working. Because it involved his personal safety, as well as the town's wellbeing, he rose to the occasion to help the FBI as well as himself. He is a middle-aged man with comfortable financial means. His wife passed away eleven years ago and his son joined the Austrian army. He is a colonel. His great-great-great father started the gunsmith business. His name was Pieter Borovenke. About 120 years later, one of the male off springs changed the last name to Borov. It was his opinion that this shorter name was easier to pronounce and therefore, it was easier to deal with his customers. Furthermore, just a minor item, it fit better on the new sign that one of the local artist created.

Ludwig Borov frequently studies his heritage and he is proud of it. After all, his, as well as most of the gunsmiths of Ferlach could trace their roots to the Netherlands, when they migrated south to a valley that they later named Ferlach. Most were peace loving and highly skilled folks, proud of their heritage. They have survived numerous major wars and kept very busy during these violent times. They manufactured guns, pistols, bayonets, swords, hand grenades and other essential items during these trying periods. They even forged arrowheads for the devoted hunters. In peacetime, the gunsmiths received orders for hunting weapons from diplomats, from Kaisers, from kings as well as from average hunters from all over the world. They were acclimated to these international exposures and they accepted it. However, they never had to deal with roboids, bats

112

and the Ebola virus. Ludwig was worried about that and he talked earlier to his friend, James Hambrush about the dangers that they are facing. It did not take them long to devise a plan for dealing with roboids. Particularly since they witnessed fatal incidences in their own town.

First, Hambrush will assume the position of police chief until they would feel confident that the roboids are gone. He contemplated to involve Johann Franzel, but decided against it; roboids might never find him. Then they will provide guards, trained in the usage of ray guns for all gunsmiths, manufacturing ray guns. Finally, Hambrush ordered a roboid from New City, programmed and especially trained for police work, aiding the humans in Ferlach. He will arrive on Monday, on a special FBI jet from new city. His name is Dactor.

On Sunday afternoon, Borov and Hambrush rearranged furniture at police headquarters. They placed the desk close to an exit door, just in case that Hambrush would have to escape. Also, he hooked a gun under his desk and he placed a few of them in a rack along the wall near his desk, in close reach. Then, Hambrush released two short time prisoners; one was smoking pot and the other drove while intoxicated. He gave them a warning and let them go. After the friends looked over the station, they nodded at each other and Hambrush said, "What do you think, Ludwig?"

"I think that we should cut out a rear exit in the back of the police station. That shouldn't take long."

"Good idea. One never knows what might happen. A rear exit is a very good idea."

After the bricklayer completed his work, Ludwig said, "I

113

think that we are ready for the intruders. All we can do now is wait for Dactor and the roboid infiltrators." James looked studiously at the floor and said, "Right. However, I do have one more thought."

"What's that?"

"I believe that we should post lookouts at the train station, at the road coming in from the north and one at the south entrance to the city. And we should keep in constant communication."

"I agree. It's a good idea."

James called two of his constables and told them to take turns watching the roads and the railroad station. He also hired two additional men, well versed with weaponry to help out. The constable watching the railroad station had the easiest job. He went inside and asked for the train arrival schedule. The train supervisor listed only four arrivals during the next two days, but no one looked like a roboid.

CHAPTER 23

On Monday, before noon, Dactor arrived in a small business jet. He landed at the Ferlach airport. Then he walked to the police station, carrying a grip with a change of clothing and when no one looked, he ran at the rate of fifty miles per hour. He walked into the police station and said in perfect Austrian dialect, "My name is Dactor. I need to recharge." James Hambrush frowned and said, "I don't have the equipment."

"Not to worry. I have all that is required on me." He opened his shirt, released two latches and flicked a miniature switch from USA to Metric. Hambrush looked puzzled, *this is a machine? He looks so real.* Dactor noted it, ignored it and said, "This a voltage conversion switch for my transformer." Hambrush realized that the output voltage in the USA is different from the one in Europe. He smiled, nodding and understanding the procedure, while Dactor plugged in. Hambrush studied the new arrival. He was a beautiful 6 foot, 4-inch-tall man, dark hair, muscular with a slim waist. He wore a new suite to fit in with the local businessmen.

"Can you talk now while you are recharging?" asked Hambrush, politely.

"Absolutely."

"Do you have a plan, Dactor?"

"No. Do you?"

"I think that I have a plan. However, I'm not in tune with roboids and their behavior."

"You do not have to be. They act like humans. However, when you recognize one, use a ray gun and shoot him through

115

his chest. This will sever his main slave connecting the main computer with the individual joint motors. If you shoot a human by mistake, unfortunately you will see blood flowing." Hambrush shook his head; *this machine has no feelings.*

"How do I know that I'm facing a roboid?"

"I can tell immediately. All I have to do is look at their cameras — ah, eyes for you."

"What is different?"

"The cameras of roboids have a distinctive reflection and darkness that humans do not have."

"I will have to watch that." Dactor smiled and when he turned, he saw a beautiful, 20-year-old, blond woman walking into the police station.

"What a beautiful human being," said Dactor to her, holding out his hand. She shook his hand and Dactor was careful not to squeeze too hard.

"This is my daughter, Erica," said Hambrush.

"I am Dactor. I will be working for you father until my creator reassigns me."

"Creator?"

"Erica, Dactor is a roboid. It's the latest design."

"I would have never believed it," said Erica, studying Dactor's body, where the legs join the trunk. Erica tried to continue the conversation. She composed herself and she finally said, "What are you doing here in this small town? Dactor noted that Erica was looking at his crotch and his sensors sent a special signal too his main computer to enlarge his private part by 50 percent. This amenity is a standard, programmed feature that every roboid possesses, male and female. Erica noticed the

116

growth of Dactor's penis. She could hardly breathe and finally said, "Let me show you the police station, Dactor." Then she turned to her father, "Father would you mind?"

"No Erica. Show him where we store the ray guns and give him one. He might need it soon."

"Yes father." Erica took Dactor by the hand. It felt soft, but in contrast, Erica's hand was hot and *she* squeezed Dactor's hand.

"Erica, you have a strong grip."

"I'm sorry. That's unintentional. But I can't help myself, looking at you."

"I believe that I understand what you mean, as a human." Dactor recalled the subprogram that human programmers install on all advanced roboids. Erica pulled Dactor to an isolated corner cell. She pulled the door closed, turned and kissed Dactor. He immediately responded and his penis increased to its full size. He grabbed her by her hips and lifted her onto a table, located next to the toilet. Then he was going to pull down her undies but he couldn't find them. She did not wear any. His program guided him to the following steps; unzip, pull out his penis and insert it into her vagina. Sensing devices detected ample lubrication in the woman's body — his sex subprogram signaled, *no additional lubrication required*. Erica had trouble controlling herself and where normally a scream would indicate her state of utter satisfaction, she simply moaned. After twelve minutes of unbelievable sexual activity, she gently pushed Dactor from her and she collapsed on the table. Dactor was confused, "I something wrong, Erica?" he asked, trying to understand the human

117

female.

"No, no. Nothing is wrong, Dactor. Everything is wonderful." Dactor accepted her statement, but programmers failed to add a closing, which would explain the human female's state of mind after tremendously satisfying experience. The self-learning chip took over and programmed a closing to a normal sexual episode. Someone knocked on the door. Quickly, Erica wiped excessive vaginal fluid from her legs and jumped off the table. She pointed at Dactor's open fly and he expeditiously squeezed his penis to remove the pressure, stuck it back into his trousers and closed his fly.

"Yes, what is it?" said Erica, still breathing heavily.

"It's your father. Is everything alright, my dear?"

"Everything is fine, dad. I am showing Dactor this high security cell, just in case he would capture a roboid alive."

"All right — good idea. When you are done with your business, come back to my office."

"Yes dad. We should be there shortly." Erica opened the door and walked quickly to the office, still recuperating from the best sexual experience that she had in her life. Hambrush looked at his daughter and he knew immediately that she engages in a sexual activity with Dactor, by observing her shaky body. He cleared his throat and said, "We are expecting two population control roboids any time now, but I don't know where they will go first. You must be on the lookout for bats, or for cages with bats in them. They might come here to take over the police department. Or they might go to Borov or to my shop."

"Mr. Hambrush, you have to stay here. Arm yourself

118

with a ray gun and a shotgun for the bats and keep your daughter with you. I will watch both businesses.

"That seems like a good plan. But, I would like to keep Erica with you. She would be safer with you."

"I am very good a handling a shotgun," said Erica, "I would use nothing less than a 12 gage. After all, I'm the daughter of one of the best gunsmiths in town and for that matter in the world." said Erica confidently. Dactor studied Hambrush and he nodded. He had no reason to disagree.

"Alright then. Let us get started, said Dactor." He already had a ray gun in his hand, but Erica rushed to get a shotgun as well as a box of 12-gage shells. Both went to Borov's gunsmith shop and Dactor introduced himself. Erica explained the rest. They turned on their cell phones to stay in constant communication.

"I am going to run to your father's business. I will be back in five minutes. You go upstairs and look down through the window. When you see something suspiciously call quickly," said Dactor.

"I will." Now all parties were in three-way communication with each other.

Hambrush sat behind the desk at the police station. He had a ray gun next to him on the desk and a shotgun leaning against the desk on his right. He was nervous. He experienced many dangerous situations. He fought the enemy in wartime and hunted lions, tigers and rhinos during peace time. However, he never faced Roboids and killer Ebola virus bats. He rose from his desk and walked to the front window, facing the small plaza. It was quiet. An elderly woman exited the bank and

119

walked to the grocery store, on the other side of the plaza. A gunsmith carried an action to a stock maker and a teen ager ran across the plaza to his school. Two tall, well-dressed persons in Austrian garb left a van, parked at the far end of the plaza. It appeared that one was a male and one a female. The male held a small cage in his left hand and he buried his right hand in his coat pocket. They stood by the van and seemed to study the downtown area of Ferlach. They had no idea that the folks in Ferlach are prepared for a roboid invasion. One pointed to the left, and the other pointed to the right. Obviously, they did not quit know which way to proceed. James immediately knew that they are strangers. Since he was in direct communication with Dactor he said, "I believe that both roboids are here. They are standing on the far end of the plaza, trying to figure out what to do.

"Do not move. I will be there in two minutes."

"Right." James looked out of the window and less than two minutes later, he heard a high-pitched sound of movement behind the police station. He turned and there was Dactor, entering through the back door, holding his ray gun pointed toward the front door, intently looking out.

"Where are they?"

"They are still on the plaza, scanning the area and getting closer."

"I will sit behind the desk. You, Mr. Hambrush sit by the clerk's desk on the other side of the office. I will know what to do."

"All right." Both were in position and they waited. In the meantime, Dactor talked to Erica and told her that he

120

believes that both roboids are on the plaza, but he remained where he was and keep on the lookout. The pair with the cage, walked around the outer periphery of the plaza and looked in all stores, arranged in a large circle. When they approached the front of the police station, they stopped. The male holding the cage, moved closer to the window and looked in. Then he turned and said something to his accomplice. Whatever the statement was, they decided to enter the police station.

"Good day," said the female, "We wonder if you could help us." Dactor looked at them and they looked at Dactor. Both knew immediately that they are roboids of the latest design, putting them into the Roboid's upper class. However, the two visitors did not know that the designer programmed Dactor for defense against hostile roboids. Knowing this advantage, Dactor said convincingly, "How can we help you?"

"We are looking for two gunsmiths that are manufacturing ray guns."

"Why is that?"

"Because, we are interested in ray guns."

"I could sell you a couple."

"We would like to buy them from the originators; you understand — the people that manufactured them."

"I understand, but that might not be possible right now."

"Why not?"

"Because their stores are closed."

"Why are they closed?"

"Because the gunsmith making ray guns was killed recently."

"Can you show us where this gunsmith is located?"

121

"This would not be possible; they are dead. By the way, what do you have in your cage?"

"That is none of your business," said the male. At this point, Hambrush grabbed his shotgun and pointed it toward the cage." The female saw that and she drew her weapon. Dactor pulled his ray gun from under the desk, aimed and fired at the roboid. Simultaneously, Hambrush fire both barrels of his 12-gage shotgun, sending the bats to smithereens. All bats died. Unfortunately, blood squirted around the office. The male roboid turned and tried to flee through the front door. Dactor rose and with one jump he reached the roboid, outside the door. He picked him up and threw him twenty feet across the plaza. The roboid still had his gun in his hand and aimed at Dactor. However, Dactor was faster. He fired his ray gun, boring a 5-inch diameter hole through his chest, disabling all circuitry. Immediately, Hambrush called the forensic team and told them what happened. They arrived fifteen minutes later, cleaned up the mess and sprayed the office with their neutralizing agent. Dactor called Lena and told her what happened and she said while flying to Tunisia, "Dactor, stay in Ferlach until we call you."

"Yes Ma'am." Erica smiled at Dactor. She was looking forward to another exiting episode.

CHAPTER 24

Lena and Julius arrived on Monday, at noon. It was an overcast and humid day. They rented a Ford and Julius drove to the Hotel Abou Nawas in Tunis. It took them forever to get there, because of a major traffic jam. They were frustrated and the air-conditioning in the Ford stopped working. Tall and good looking police officers tried to determine the cause of the accident in order to charge the guilty person. Unable to do that, they finally separated the parties involved in the accident and moved them to a safe area. Then they cleaned up the debris and ordered the drivers of the stranded cars to move. Julius observed and turned to Lena, "Something does not seem right. It seems that the police officers are totally lost."

"I noticed that. I bet that they worked for Christopher Barony."

"That's what I was thinking. Could it be that the whole police force is already infiltrated by roboids?"

"It almost seems that way." Finally, they were on their way. When they arrived at their hotel, both were tired and they needed a shower. Then they went to the restaurant below and ordered two bottles of imported, dark beer. Surprisingly, it was cold. They also had a quick, late lunch; ham sandwiches, American style. Julius put down his bottle thoughtfully, and said, "Are we going to see Christopher Barony?"

"Yes, it would be a good idea. I am wondering how we should approach him and what questions we should ask."

"Good questions. I'm not sure that he knows that we are on to him."

123

"Should we call him and make an appointment?" said Lena.

"I would prefer to catch him by surprise. Actually, his office is just past the Medina. Besides, FBI agents normally do not make appointments."

"That's true. Let's go see him." Julius packed the Ford with their usual weaponry and he drove carefully, but even then, he barely missed an electric trolley. The conductor blew his annoying horn at him and Julius exhibited mature control avoiding road rage. Lena pressed her lips and decided not to get involved. He passed two large palm trees and right past it, was the police station. It was a large building, with numerous windows, curved on top. A painter was just finishing applying white wash to the brick building. Julius parked the Ford in front. Both agents left the car, checked their ray guns, and entered the building. A tall man in uniform stopped them and said, "What is your business?" Lena looked at him and said, "I believe that I have seen you before."

"That is impossible. I never leave the police station."

"I see. Is it possible that you might have a brother?"

"That is impossible. I have no brother or sister." Then the man's eyes flickered. "I ask you again, what is your business here?" Julius decided to answer, "We are here to see your police chief."

"Why do you want to see the police chief?" Lena became irritated, decided to step in and said, "We might be looking for a job. Can you help us?" The officer hesitated, then said, "I am not trained for hiring people."

"Why not?

124

"I am not trained for this type of thing."

"Why do you keep repeating yourself, officer?"

"I am not repeating myself."

"Oh, yes you are repeating yourself. Only robots keep repeating themselves. Are you a type 3 low-life robot, or are you a stupid human?" The officer certainly does not want to be a type 3 robot and neither does he want to be a stupid human. His head turned in a jerky motion, looking for a response and he finally assimilated, "I will allow you to see Chief Barony. Walk down the hall. His office it the last one on the right."

"Thank you very much," said Lena, grinning. Julius shook his head, "That was good. I just now learned something."

"What did you learn, Julius."

"How to confuse a roboid's reasoning power." The door to Barony's office was open. Julius stepped in and said, "Greetings, Chief. This is Lena McCabe and I am Julius Hersey. We are FBI agents and we flew in from DC." Barony was completely taken by surprise and he said, "How did you get past my guard?"

"That was simple, Chief. We confused him and then he told us to see you." His eyes flickered, "I see. What can I do for you?" Lena decided to start, but she still pondered how to begin. Then she decided that she would act as if she knew nothing about roboids and said, "Our supervisor sent us down here to investigate the death of an unusually large number of people. What do *you* know about it?"

"People are dying all the time. What exactly are you referring to?"

"The FBI received a report that nearly all people in El

Habibia died a horrible death. That is over ten thousand folks."

"Yes. I have heard about that also."

"Are you investigating what happened down there?"

"No, not right now."

"Isn't El Habibia a part of your jurisdiction, Chief?"

"I suppose it is. Soon, I will send someone down to investigate." Lena realized that this line of interrogation will get her nowhere and she decided to try a different approach, "We visited the town not too long ago and we saw a man cleaning up the whole town."

"Really."

"Yea, really. He told us that his name is Peter and that he works for you. What do you know about that?"

"Right. Peter works for me. The whole Tunisian police force works for me, including drivers, cleaners, genitors and numerous other skilled and non-skilled people? How do you expect me to know the names of all these folks?"

"If that's the case, how come is it that you remember Peter?"

"You have a point there," said Barony.

"That is a very strange response for a police chief. Did you learn to say this response?"

"Yes, I did." Barony frowned and thought, *did I make a mistake?*

"Do you memorize all responses when you are talking to people?" Now Barony became 'suspicious' and he acted busy shuffling papers. He said nothing and gave Lena a strange look. Lena continued, "When did you start in Tunis as police chief?"

"About two months ago. Why do you ask?"

126

"We are curious. Where did you work before?" Barony stood and his seat slid back, nearly falling over. Then he said, "That is none of your business. I believe that the meeting is over."

"Not so fast," said Julius. I have a couple of questions." Barony sat again and said, "Make it quick. I am a busy man."

"I can see that. Do you know what killed the population in El Habibia?"

"No, I do not."

"Why not. Is it not your responsibility to find out what killed them and report this to the Tunisian President?"

"I am not in contact with the president." Julius hesitated and decided to dig deeper, "When is your next shipment of men arriving?" Barony stood again and this time the chair behind him tilted over, landing on its side, "This meeting is definitely over, agents. I suggest that you go back to the United States and allow us to run our business." Julius and Lena rose and headed toward the door.

"Thank you for your time Chief Barony. It was nice talking to you."

"Same here." Barony followed the agents to the door and when they cleared it he called his guard and slammed the door. Julius and Lena returned to their hotel. While Julius drove, Lena said, "Is it possible that Barony is in charge of the whole population control operation?"

"It's possible. Someone must be in charge. He seems to be a most advanced roboid, but he still has many short comings."

"I agree. We have to learn more about their plans and

127

soon we will have to initiate roboid control before they *really* take over the world.

"God. That would be a catastrophe."

"That would be the end of the world, as we know it."

"That would be worse than being invaded by aliens."

"I agree."

CHAPTER 25

Managing the roboid manufacturing operation was fun for Pharos. Fun? Yes, it was *fun* for him. Now he has learned to differentiate between fun and boredom. That means that he also developed a sense for time. His self-learning program keeps evolving exponentially. When he works with something that is fun, times moves fast. And when he has to shuffle papers on his desk, times passes slowly. However, mostly, time goes fast for him. Not only must he maintain a constant production flow, producing male and female roboid parts and assembling them, but he also must be extremely observant of his surroundings and of Antony Marino and his wife Lexia. Occasionally he has visitors. Usually they deliver roboids for repair. Another important aspect of his job is to repair and upgrade roboids. These are roboids that are mostly programmed for domestic service, manufactured by the Bergen Company and other roboid manufacturers. He upgrades programs, installs vocal cords and vaginas, made from newer and more flexible materials. For males, he installs new, flexible and more durable penises, with improved expansion capabilities. Frequently, he also changes the facial skin covering. These are only a few of the repair work categories for roboids. Marino keeps close track of them, because someday he might have to use them and reprogram them wirelessly for other duties, based on a need basis. That is one of the features of roboids that owners do *not* know about.

Pharos stood by his desk and looked at the calendar, hanging from the wall. It is a habit that he developed recently, imitating humans. He doesn't need to do that, because he has a built in time clock, upgrading every second of the day. Today is Tuesday, still early in the morning. Krista is on the dock below, loading four roboids, programmed for population control. She had trouble seating them correctly and finally strapped them down. The truck will take them to the Dallas airport and a private plane will fly them to Tunis. Pharos just now remembered that on this day, Krista will have to perform another important job. He rushed from his office to the dock area below, where he found Krista. She looked at him and said, "Pharos are you checking up on me?" Pharos awkwardly shook his head and said, "No. But I do have new instructions for you."

"What are they?"

"I want you to make another shipment to Ferlach, Austria, as soon as possible."

"How many roboids, Pharos?"

"I want you to load three male roboids, programmed for population control and one female, programmed to perform multiple functions."

"No problem. I still have another truck waiting. Or should I load all eight on the same truck?"

"Load all eight on the same truck, since they all are going to the same airport. But be careful that you do not get them mixed up."

"I will not get them mixed up, Pharos."

"Good." Satisfied that he has good control over Marino's latest instructions and that he did not forget any of the details, he decided to take a tour through the plant. He started in an area, where they had a recent problem. He watched automatic equipment form, copy and machine titanium femurs.

130

They are categorized and stamped according to the roboid's height. Usually male femurs range from fourteen to seventeen inches in length, whereas female femurs are usually up to four inches shorter. A few days ago, a robot tested a male roboid before final shipment. He walked with an unusually large limp. An automatic rejection program issued a rejection; a red light flashed and a siren blew, drawing both Antony and Pharos to the inspection station. Both saw the roboids laying on its side. They determined that the automatic parts identifier mistakenly marked a femur as female instead of male. This resulted in a limp, tilting the roboid beyond its center of gravity. Pharos returned the limping roboid to the repair station and Antony said, "Find out what caused this misidentification and repair it."

"Absolute," said Pharos, while reviewing the program. Since then he has taken tours through the plant to make sure that body parts are identified correctly.

Pharos returned to his office and decided that it is time to place a phone call to Lena. She answered after the fourth ring, "Yes, Pharos. What is it?"

"I am reporting to you, Lena."

"In the middle of the night." Pharos looked at the clock on the wall, "Sorry. I did not realize that you are sleeping."

"Humans usually sleep at night. What's new at the roboid factory?"

A truck is headed for the airport, as we speak. Four roboids are on it, programmed for population control. They should arrive in Tunis tomorrow."

"We will be waiting. How about bats, Pharos?" Do you have bats in your factory?"

131

"Absolutely not. We have no bats here. Do *you* know who has the bats, Lena?"

"We believe that the police chief in Tunis is in charge of the bats. He must still have a cage full of them."

"I see."

"What else is new?"

"Oh, I almost forgot."

"You forgot. Don't tell me that you are acquiring more human traits."

"Sometimes I wonder about that. What I have to tell you is very important, Lena. On the same truck I have an additional four roboids. These are going to Ferlach, Austria. It will consist of three male roboids, programmed for population control and one female roboid, programmed for multiple functions. Her name it Tania and her human name is Angela Barret. They should arrive in Ferlach next Monday.

"Monday? Wow. That won't give us much time. Fortunately, we still have the FBI jet. Right now it's the fastest jet in existence."

"Yes, I know." After a long pause, Julius looked at Lena, next to him, "What's happening? Did Pharos hang up?" Lena shrugged her shoulders, "I don't think so. Pharos, are you still there?"

"Yes, I am. Why do you ask? Is something wrong?"

"No. nothing is wrong, Pharos. By the way, I will call the Mayor of Ferlach and tell him that four roboids will be arriving."

"Good. If I have more news, I will call you.'"

"Make sure that you report all of Marino's plans and

132

movements"

"I will." Both disconnected.

CHAPTER 26

After Lena took a shower, she called the Tunis airport and inquired when the next plane from Dallas will arrive. A ticket agent looked it up and told her, eleven twenty.

"We don't have much time, Julius." Lena talked through the bathroom door, hoping Julius will hear her. He opened the door, towel wrapped around him, "I heard you dear. I won't be long. Could you order breakfast? We still have time for that."

"Yes, why not." By the time Julius was dressed, a waiter knocked on the door with a cart, breakfast steaming. After breakfast, the agents went to the Ford and Julius drove to the airport, waiting for the plane to arrive. It was a smaller plane, only ten minutes behind schedule. First, two businessmen left the train. Then a young couple stepped of the gangway. They looked in both directions, trying to figure out which way to go. Finally, the female pointed down the corridor and both left in that direction. Next, four tall, very good looking men slowly turned right and headed toward the main exit. Normally folks look for a baggage conveyor, but not these four men. They had no baggage. Lena pocked Julius, "Let's go to the Ford. We may not have much time."

"Right." The agents rushed to the parking lot and Julius drove the Ford to the main entrance.

"We are in luck. They are just now stepping into a taxi. Let's follow them."

Right," said Julius. Following the taxi was easy. The driver twisted and turned through the narrow alleys. The roboids were unaware that they might be followed. That was not part of their program. However, they looked in both directions, probably photographing the surroundings and storing it on one of their memory chips. The taxi driver stopped in front of the police station, while Julius parked the Ford nearby and out of sight. Christopher Barony waited for them in front. He inserted a memory stick into the chest of the first roboid and then removed it. Barony repeated the procedure three times with the other roboids.

"Barony is reprogramming them. I wonder what for?" said Julius, concerned.

"I'm not sure. I have the feeling that we will find out soon." Another police officer came out, carrying a cage; it was covered with a cloth.

"I know him. Isn't he the officer that blocked us when we tried to see Barony?" said Lena.

"Yes it is. No question about it. I bet that he has bats in the cage." As if directed by an invisible force, the officer passed the bats on to one of the four roboids.

"You are right. We must try to stop them from freeing the bats. That could be a catastrophe, including for us." Next, the agents saw a sequence of strange behavior. Barony and the other officer returned to the police station and left the four roboids standing in front. After a few moments, one roboid pointed in one direction, while another pointed in the opposite direction. The third roboid, with the cage, lifted his hand and pointed west. Then he started walking. Eventually, the other three followed.

"We better hurry and go back to the Ford." Julius started the Ford and slowly followed the roboids. They continued to walk faster and faster and when they cleared the city, they continued on Highway P7. They stopped for a moment, looked in all directions, then they started to run. In order to keep up with the roboids, Julius was forced to drive between fifty and sixty miles per hour on a sand-covered highway, in dire need of repair. The roboids continued to run in a westerly direction. One decided to look back. Julius stepped on the brakes with all the force the he could muster with his foot, creating a blanket of dust. The roboids stopped and now all looked back, probably focusing in on the truck. However, instead on attacking the truck, they decided to run faster. Now they ran at seventy miles per hour. Lena said, "Typical roboid behavior. I guess they are not programmed to attack right now. By the way, this must be their top speed."

"It seems that way." Lena looked at her iPad and said, "Those bastards are going to El Habibia. What the hell are they going to do there?" Sure enough, after running for about fourteen miles they turned south and headed straight to the long building that Peter got through cleaning just three days ago. They walked in through the main door and left it open. Julius stopped the Ford, parked it near the building, facing the windowless wall, and walked to the front door.

"Do you have the ray gun?" said Lena, worried.

"Of course. I also have my pistol and a shot gun pistol, just in case that I have to shoot at bats." Both walked down the hallway, ray guns pointing ahead of them. When Julius passed the first door on the right, he looked in and saw the four roboids, checking the cage. Nervously he motioned to Lena to stop. Both agents carefully looked it. The roboids paid no

136

attention to the humans. Then one covered the cage with the cloth. One roboid reached to the floor and lifted one corner of a six foot by six-foot steel cover. It must weigh at least one thousand pounds. Another roboid place the cage in an opening below. Then the roboid, holding the steel cover slowly placed it over the opening.

"We could never get the bats out of the cage, unless we have a front end loader or a lift truck to remove the plate.

"You are very observant Julius." Julius shook his head, "Women."

Next, a roboid looked for an electrical outlet. When he found one, he plugged in a surge protector and then all the roboids plugged themselves into the surge protector. Lena grabbed Julius's arm and said, "They are recharging. That should take about two hours. Julius looked at Lena and said, "This would be a perfect time to kill them."

"I agree. But if w4 did that we wouldn't find out what they are planning."

"You are right. Let's grab a bite to eat and return. We must not lose track of the roboids."

"Good idea." The agents rushed to the Ford and Julius drove nearly one hundred miles per hour to the outskirt of Tunis and ordered enough takeout food and drinks to last them for a couple of days. They ate and drank, while they returned. Julius parked the truck in the same place. Lena left the truck and went in the front door and checked the room. The roboids were still in their powered down state, standing around the surge protector. She returned to the truck, sat in the passenger seat and said, "They are still powered down. Get some sleep, Julius. I

137

will watch the building."

"You should get some sleep."

"I can't. I'm all wound up." Julius frowned, stretched behind the wheel and closed his eyes. One minute later Julius snored, while Lena watched him, *what a beautiful man*. Finally, Lena also dozed off for a few minutes.

CHAPTER 27

Lena woke up. She jumped while straightening and thought, *dammed, I fell asleep. If I screwed this up, I will never forgive myself.* Again, she left the Ford truck and tiptoed to the room where she saw the roboids last. She was relieved when she saw them still standing in the same position. When she returned to the truck, Julius was awake, "Is something wrong? Are the roboids still there?"

"Yes, Julius, they are still there. I dozed off and I thought that they might have left."

"What time is it?" Lena looked at her watch. She illuminated the dial, "Five thirty." Julius stretched and stepped out of the truck. He walked back and forth to loosen his limbs. Then he returned to the truck and reached for a bottle of water. He offered it to Lena. She drank almost half and gave the remainder to Julius. He drank the other half. He went behind the truck, looked in all directions and relieved himself. Lena noticed and said, "I have to go too."

"Go behind the truck. No one is in sight." Lena followed Julius' instructions.

"Boy, now I feel better. I am wondering if the toilets in the building work," said Lena.

"Why should they. Roboids don't need them." Before Lena had an opportunity to respond, two roboids left the building.

"What should we do, Julius?"

"We should follow the roboids."

"What about the other two?"

139

"We have to take this chance. We have only one car."

"That's true." The roboids walked fast, and headed straight north on the same road that Julius and Lena took before. Then the roboids ran fast and they left a cloud of dust behind them. Julius revved up his truck, trying to keep up with them. Lena was confused and she said, "What about the bats? They didn't take their cage. Did they leave the bats behind?"

"How should I know, Lena. But, I did see one of the roboids carry a small plastic bag. Perhaps they only took only one or two bats with them. They are lethal and they could wipe out a whole neighborhood in no time."

That's scary. Now I wonder what the other two roboids are doing." Lena thought and said, "Just think, if the four would split and everyone wound go to a different town, they could wipe out your town at once."

"That's a scary thought." Now Julius is driving at sixty-five miles per hour again and he is keeping up with the roboids. Unexpectedly, they took a narrow street, headed in a north easterly direction. Lena immediately checked with her iPad.

"What did you find," said Julius, impatiently.

"I will tell you if you give me a chance. Apparently they are headed to a town called, Qued Ellil. It's much larger than El Habibia."

"Is it a metropolis?"

"It might be. The population is about 60,000 to 70,000."

"Shit. That's all we need. These fucking roboids definitely have a plan. First they test their theory on a small city and wipe it out. Now they are in the process of killing all the folks in a larger city. We must put a stop to this frenzy, before

140

the really take over the world."

"I'm sure that's their ultimate plan." The two roboids arrived at the outskirts of town. They ran into a side street and for a moment Julius lost them. He floored the gas petal, sand spitting from the rear tires, and followed them. When he turned around the building ahead of them, he nearly ran them over. One of the roboids held a plastic bag above him and both agents saw that he released two bats.

"Shit, that's all we need." Julius hit the brakes. The truck buckled and both agent stepped out, holding their ray guns in front of them. The Roboids heard the noise, turned and both were ready to shoot their weapons. One roboid did shoot, but missed Julius by inches. This time Julius and Lena were faster. They fired their ray guns, with the help of their laser aiming devices, hitting the roboids in the chest. They stopped moving, but both remained standing. Lena looked for the bats. She saw them flying over the near building. Julius ran around the building, drawing his shot gun pistol. He was ready to fire as soon as he would see the bats. The bats were nowhere in sight. It was still early in the morning and the agents looked in different directions. They were at a loss and Lena said, "Where the hell are they?"

"I wish I knew."

"Just be careful that one of them doesn't land on you. If she does, you have less than one hour to live."

"Do you suppose that they might have found a hole and crawl in it?"

"Why do you say that?"

"Most bats sleep during the day."

141

"That's right. I forgot about that," said Lena, pouting. Slowly the agents walked down the street. Then they returned to their truck and sat in it. The two roboids still stood motionless in the middle of the road and Julius said, "Let's look for the police station." Lena pulled out her iPad, texted something and said, I found the police station, Julius. Drive back to the main road. It leads directly to the police building."

"Right." Qued Ellil is a fairly big town and it took Julius twenty minutes to reach their destination. On their way they saw beautiful, white washed and tiled buildings, a mosque and luxury homes for the rich. Julius parked in front of the police station and the agents walked in. A sergeant sat by the front desk and spoke to them in Tunisian.

"Sorry sergeant, we do not understand."

"Ah. Americans – that figures; they expect everyone to speak English."

"Exactly, said Julius. We are FBI agents. I am Julius Hersey and this is Lena McCabe. We travel to Germany, to Italy, to China, to Japan and a few other lesser countries, like Tunisia. Do you expect us to learn all these languages? Normally we take an interpreter if we need one. No time today. This is an emergency. Get the captain or someone in charge."

"That is me. I am in charge. The captain is on vacation. I am Sergeant Karim Jomas. What can I do for you?" Julius looked at Lena and both new that working with this person will not be easy. Lena decided to start, "Put out your arm, sergeant."

"What for?"

"Please, don't argue with me. Put out your arm." Finally, the sergeant put out his arm in the direction of Lena.

She had the ring with the needle on her middle finger. She reached for his arm and squeezed it.

"Ouch. What the fuck are you trying to do to me?"

"I see, all of a sudden, your English is better than average. I had to verify that you are a human."

"What the hell are you talking about, human?"

"Have you heard about the catastrophe in El Habibia?"

"Yes, I did. What does that have to do with us?" Lena reached to Julius for help.

"Let me explain to the sergeant what's going on." Julius spent nearly one hour, starting with their first trip to El Habibia and ended with the lost bats in Qued Ellil. When he was done the sergeant said, "I'll be god dammed. I had no idea were in such trouble. What do you want me to do?"

"We have to isolate your city and the area where the bats appear to be. Then we must get a medical team down there. They must take a 0.5 percent sodium hypochlorite solution to disinfect people and objects."

"Where the hell would I get this chemical?"

"Let the medics worry about that. They might have to go to Tunis for that." Julius continued, "Then we have to educate your people to look out for bats and stay the hell away from them. Take some of your best officers and arm them with shotguns."

"How do we look for bats?" asked a disillusioned and worried Karim Jomas.

"That's a very good question. Get someone from the medical team and buy a powerful sound emission device with the capability to emit sonic sound emissions of various

143

frequencies. That will draw them out during the day. Your officers should be ready with their shotguns and when they see the bats, shoot them. There might be more than two, though two infected ones are what we are looking for."

"That easy?"

"It won't be that easy, but in might just save your town?"

CHAPTER 28

Sergeant Karim Jomas finally realized what he faced in his town. If he didn't jump into action immediately, he might lose all the folks in his town as well as the dogs and cats in just few days. And that might include him. He knows what happened in El Habibia. He heard how people died. He saw pictures of blood soaked up by the sand, white skeletons, and he thought, *boy I am going to stay away from this catastrophe.* The sergeant took three trucks loaded with officers and supplies, as well as a front end loader to gather up the two roboids, standing in the street. Under the direction of Julius and Lena they drove south. When they arrived, the agents finally had a chance to study this area of town. It is the underprivileged end of Qued Ellil. The roads need repair and the walls of the building began to crumble. Tunisians were passing time, some sitting around tables in front of public places, drinking their favorite drink or playing a game. When they saw the truck driving by, they stood and wondered what's going on. Some followed the slow moving trucks. An officer jumped off the truck and guided the locals back to where they came from. One resisted and he forced the officer to get heavy-handed. Another officer said something to the rebel and when he heard what he said, all of the locals jumped up and ran the other way.

"I guess he made his point," said Lena.

"Yes, he did. I don't think that we have trouble with the other locals." Julius drove his truck to the street where he and Lena left the roboids. They were still there, standing in the same place. The ray guns must have ignited the burnable material inside, because only the titanium skeleton was left, but a few parts still smoldered. Julius left the truck and looked for the bats. He checked his watch – 4.30 p. m. He walked down the next alley. What he saw turned his stomach. Three locals still sat around a table spitting blood. The sand was colored red. Slowly they degenerated. The sergeant joined the agents. He couldn't believe what he saw and said, "People told me about this bug. I had a hard time believing it. But now I can see it myself. Mr. Hersey, what do you want me to do?" Hersey conferred with Lena. Then he turned to the sergeant, "We must isolate the whole area. Do not let anyone leave and do not let anyone enter."

"How, Mr. Hersey? How?"

"Use your yellow policer tape. And position your policemen all around in intervals"

"Right." While the officers isolated the area, Lena and Julius checked other alley. Most presented a similar picture of death and despair. Julius gave another order to the closest officer, "Have your shotguns ready and shoot all the bats that you see. Spread the word around the perimeter."

"Yes sir. What are you going to do, sir?"

"I am going to slowly start the sound emission device. Be ready to kill the bats."

"Yes, sir," said the officer and he ran around the perimeter informing everyone of Julius' intent. Julius went to his truck, opened the rear gate, hooked up the battery to the device and he slowly started to crank up the device. He and Lena also had their shotgun pistols read, in case a bat might drift in their direction. Every one heard the high-pitched sound.

146

They had their shotguns ready to shoot. At first nothing happened. But Julius intentionally started slowly. He did not want too many bats at one time fleeing from their hideouts. He had no way of knowing if bats existed in this area before the roboids released their two bats. Julius increased the sound from the device. On the other end of the street, a shotgun went off. The sergeant called on his cell phone. He listened, then he said, "Mr. Hersey, one of my men killed a bat. But he does not know if it was the roboid bat."

"Better than nothing. At least I know that my sound emission device works." Now Julius cranked faster and longer. A few seconds later three shotguns went off. Someone called the sergeant, "Sir, we killed another two bats." The sergeant shouted in his cell phone, "Keep watching." And Julius again cranked faster. Lena looked up and she saw two bats nervously circling as if they lost their orientation and they hit a wall. Lena aimed her shotgun pistol at the bat and pulled the trigger. She missed. Then she shot again. This time she hit the bat, and another officer killed the second bat. Now Julius emitted his provoking sound continuously for over five minutes. Then he quit. An eerie silence set in and no one saw additional bats flying around. The sergeant looked at Julius and said, "Would it be really possible that we killed all the bats in the neighborhood.

"Very possible," said Julius, relieved. The sergeant nodded and said, "What should we do now?"

"Form a straight line and comb the area. Touch nothing. Save the healthy folks and disinfect the dead ones."

"What are you going to do?"

147

"We have to fly to Ferlach, Austria. We have a similar emergency there. Right Lena?"

"Absolutely. What are we going to do with the other two roboids and the cage with bats in it?" The sergeant listened, "There are more roboids around here?"

"Yes, sergeant. They are in El Habibia. And guess what, the whole police force in Tunis is made up of roboids."

"Holly shit. I cannot handle all this. This is way over my head, Mr. Hersey." Julius and Lena stepped aside, talked and Lena turned to the Sergeant and said, "We are going to give you help. He will arrive sometime during the next two days. Wait for him at the police station. He is a friendly roboid and his name is Dactor. He works for us and you can trust him. He is programmed as a roboid defender. We will add a subprogram to his mother board which will allow him to assume a leadership position here in Tunis."

"Are you sure that this will work, Ms. McCabe?"

"I am positive." Lena made a phone call to Police Chief Hambrush. She told him that Dactor is being reassigned. She transferred wireless instructions to Dactor: his new assignment, the location of this assignment and the means of his transportation. Then she looked at the sergeant, "It's all set." The sergeant just shook his head. He was speechless. Then he found one of his officers and ordered him to operate the front end loader, pick up the roboid scraps and load them on one of their trucks. Julius and Lena went to the Ford truck and left for the Tunis airport.

CHAPTER 29

Dactor was powered down, standing in a corner, at the Ferlach police station, when Lena upgraded his status to leader. While the new programs transferred into one of his miniature hard drives, with all his new instructions, he returned to full awareness. He changed clothes, washed the dirty ones in a washing machine and packed everything that he owns. As he zipped up his bag, a voice behind him said, "What are you doing Dactor?" He looked where this soft and pleasant voice came from. He saw that is was Erica.

"I am reassigned to Tunisia."

"Oh no. I will miss you so much, Dactor, said Erica with a crying voice." Dactor searched for an appropriate response in his vast and complicated network of information, while his eyes slightly flickered. Finally, he said, "Erica, I am sorry. But I must follow the directions of my program." Erica looked at him, realizing that, after all, this is just a machine, and she thought, *but his sex appeared so real and so satisfying.* She kissed him on his forehead and before she left she said, "Good luck to you." She knows that no one in her future life will ever compare to Dactor, though she might get pregnant sometime in the future, perhaps even without experiencing orgasms. Dactor stopped in his tracks, listening to another message, "You plane is ready at the Klagenfurt airport."

"I will be there in less than twenty minutes," was his response. Dactor wished Chief Hambrush well and he stepped outside the police station. James Hambrush watched while Dactor headed toward Klagenfurt, along the long stretch of road way pointing out of town. In just five seconds he increased his velocity to about one hundred miles per hour and ten seconds later Dactor was out of sight, heading for the plane. While on the FBI jet, Dactor collected implements of destruction to prepare himself for the confrontation with the roboids, programmed for population control.

"I am now cruising over Qued Ellil," said the pilot. Dactor looked out the port hole and said, "Cruise at three thousand feet elevation and at two hundred miles per hour."

"Affirmative." He opened the side door of the FBI jet and threw out his implements of destruction. The giant parachute opened and the package drifted toward Qued Ellil. Then, Dactor stepped on a two-inch platform, outside of the jet, closed the side door and jumped, arms stretched out pointed toward the parachute. In a few seconds, he caught up with the package and held on. He guided the chute toward the center of town. When he was within fifteen feet of the ground, he let go and landed safely on his feet. The package dropped right next to him. Sergeant Karim Jomas stood in front of the police station and said, "How the hell did you do that?"

"Did what," said Dactor and continued, "Let us see what we can do." Watching Dactor's feat, the sergeant was still recuperating from shock.

"Let us go into your office and review what we have for resources," said Dactor, taking the lead. He stopped by the sergeant's desk and said, "Have you been back where the bats were last?"

"Yes, twice. We picked up all the bones that we could find and burned them. Then we disinfected the whole area. We

150

temporarily relocated all the people that remained alive."

"Good. Let's go there and if we do not find any more bones, it should be safe to move the people back." Speedily, Dactor checked every house in the affected area and concluded that it is safe to return the remaining folks to their habitat. Of the three hundred fifty people that lived there, only twenty-five remained and only four are complete families. When sergeant Jomas reviewed these statistics, he realized the effectiveness of these killer bats that carried the virus. He was convinced that an all-out war was in order and no measure would be too lenient in preventing another catastrophe. Unfortunately, he only could relate this analogy to his small domain. With this in mind, he said, "Dactor, what do we do next?" Dactor knew exactly what to do and said, "We must go to El Habibia and find the other two roboids." The sergeant packed up two trucks with weaponry and men and headed toward El Habibia. Dactor ran ahead, carrying his ray gun and his shotgun. When he arrived at near the town periphery, he stopped for a moment and searched for the database that contained the information of previous proceedings, in relation to this town. And there was plenty. He found the long building where the roboids last stayed. He checked it and it was empty. He called the sergeant and told him to stay where he was. Dactor checked every room, including the room where Julius and Lena saw the roboids last. It was also empty. The manhole cover was still in place and it was an easy task for Dactor to remove it. He looked in, *didn't Lena program that bats should be in there?* He reviewed the program, *yes there should be bats down there.* But the hole was empty. Dactor left it open and he faced an impasse. He jumped out of

151

the hole and walked out of the building. Then he saw the sergeant with his truck standing about twenty feet from the building, *didn't I tell the sergeant to wait at the edge of town? Yes, I did. Well it is too late now. What should I do next?* Dactor down loaded more information, Peter cleaned up the town just a few days ago, *could it be that Peter is still in town, finishing the cleanup work?*

"Sergeant, check every building in El Habibia for a friendly roboid. He might still be here cleaning the buildings. His name is Peter," said Dactor, impatiently.

"Absolutely." The sergeant directed his police to search the town. On the far end of town, one officer found a roboid standing in a corner, powered down. He called Dactor and told him about it. Dactor arrived and confronted him, "What is your name?" The roboid powered up, evidenced by a slight vibration of his structure. Then he said, "My name is Peter. What can I do for you?"

"What are you doing here?"

"I am cleaning up the town."

"If you are cleaning up the town, why are you powered down?"

"I have completed my task and I have no new instructions from my supervisor."

"Who is your supervisor?"

"My supervisor is Christopher Barony. He is the police chief of Tunis."

"If you are done, why did you not report to Christopher Barony?"

"To report is not a part of my program."

152

"What is your program?"

"This time my program is simple; to clean up the town."

"I see. I think that you should work for me, Peter?"

"Would that be possible?"

"Of course that would be possible, because I am programmed as a leader."

"What is your name, leader?"

"My name is Dactor."

"Go ahead, Dactor. I will allow you to perform a reprogram." Dactor did not waste any time. With Peter available to him, he instinctively formed a plan to deal with Christopher Barony, and said, "I am reprogramming you for undercover work. You will be taking my orders."

"What will I be doing?"

"You will go to the Tunis police station and tell Barony that you were responsible to deliver bats to David Barkley. When he asks you where you got the bats, you tell him that you were not programmed for that answer. And when he asks you about Peter, you tell him that Peter was destroyed by a policeman from the Qued Ellil police department. You go that?"

"Yes Dactor, I go that."

"Does that mean that I will have a new name, Dactor?"

"Yes, you will have a new name. As of now, your new name is Baruk." Dactor opened the chest plate on Baruk's chest and made alterations on his chest plate. The most important was that he escaped from the roboid farm in New City, operated by Ramon Bergen. Then he flipped open a machine language keyboard and reprogrammed Baruk. When he closed the chest plate a surge of renewed energy travelled through the millions

153

of electronic paths in Baruk's vast signaling network. Dactor was satisfied when he saw the visible change in Baruk's appearance. Dactor nodded, a human trait, and said, "Baruk what happened to the two roboids that were in the main building?"

"Christopher Barony recalled them."

"I see. And what happened to the cage with the bats?"

"I believe that they took them along." Dactor had all the information that he needed for his plan and he said, "Baruk, go to Barony and tell him that you need a home. I am sure that he will accommodate you. But do not forget that you are working for me."

"Right." With that, Baruk left and ran to the police station of Tunis.

CHAPTER 30

Julius and Lena slept on the FBI jet. Lately, they haven't had too much sleep, chasing roboids and their killer bats, carrying the Yambuku DC Ebola virus. While their jet entered south European air space, they had breakfast on board. The steward made scrambled eggs, hash browns, and sausages, with orange juice and Columbian coffee. When they were done, Julius said, "Boy that was delicious." Lena agreed. They still had time to clean up and change clothes. Just as they had their grips repacked the pilot landed the plane. This jet does not require a very long runway, consequently the pilot landed it on the Klagenfurt Airport. On the way to the rental agency, a snow plow removed enough snow to keep a narrow walkway clear. And again Julius rented transportation. This time he ended up with a Mitsubishi truck. They needed it to carry all their weaponry securely. It took Julius just twenty minutes to reach Ferlach. He remembered where the police station was and he walked in, followed by Lena. James Hambrush saw them first, "Hello. How are the FBI agents for the USA?" Julius and Lena smiled and Lena decided to answer, "Tired and not enough sleep. We just flew in from Tunis."

"Did you get all the roboids?"

"Most of them. We reprogrammed Dactor to be a leader. It will be his job to deal with the roboids in Tunis."

"He did a wonderful job, here in Ferlach. Do you think that he can handle it?"

"Dactor can do the job." Hambrush thought about Dactor and in conjunction with this he thought about his

155

daughter Erica. He saw her crying, right after he left. Hambrush thought, *what is this, my daughter is in love with a machine?* However, he quickly stripped these thoughts out of his mind. After all, when he first met him, he assumed that he was human. Yes, roboids are wonderful machines, as long as they are friendly. This brought Hambrush back to reality and he said, "Did you check in to the hotel yet?"

"No, we didn't. We should do that, before the roboids arrive. Oh, do you know if Hans Weider is still in town?" asked Lena.

"Yes, he is. This year he did not go to Boyne Mountain in Michigan to be a ski instructor."

"He did not?"

"No. He still works at the hotel and he is giving ski instruction right here."

"Good. We are going there now to check in."

"I will see you later. If the roboids should arrive, I will call you immediately." Julius and Lena returned to the truck and drove to the hotel. It was just across the plaza. Julius parked in front and when he stepped out, Hans ran out and greeted them.

"So, you decided to stay in town, this winter?"

"Yes I did. I hoped that you would return and make me another offer." Lena nodded, "We were hoping that you would be available, Hans."

"I am ready. What do you need, Ms. McCabe?"

"First we have to check in then we will have one of your delicious sandwiches. Can you join us?"

"Absolutely. I am ready for a cup of coffee." Julius and Lena finished their sandwiches, while Hans Weider finished his coffee. He looked a Lena and she was obliged to ask him, "So, have you heard anything new while we were in Tunisia?"

"No. Nothing new. From what I remember, we should expect four roboids."

"Yes. My sources tell me that they should arrive today, and they are programmed to kill anyone that produces ray guns."

"Right. Nothing has changed since the last time that we talked. Presently, we have only one gunsmith left in town that produces ray guns. His name is Johann Franzel," said Hans Weider, pressing his lips.

"Correct. That is what we have to concentrate on." Lena stood and went to the desk, "Put the bill on my tab. We will pay before we leave."

"That is not a problem ma'am." All three went to the Mitsubishi and Julius drove it to the police station. He took out a black, large and heavy bag. Hans took it, "I will carry that." They walked in and Hambrush looked up, "I see you found Hans." Julius nodded. He took out three ray guns and three shot gun pistols and said, "These are just in case that the roboids decide to bring bats. One never knows."

"Good thinking said Hambrush. I am ready. I haven't changed anything since the last time. It worked perfectly." Lena sat on the corner of a desk and looked at Hans and said, "You and I should drive to the Johann Franzel gun store, while, Julius should stay here at the station."

"What do you think James?"

"That sounds like a good idea. Hans knows where Franzel lives. Armed with their weapons, Hans volunteered to

157

drive to the Franzel gun store, while Lena enjoyed the view. His store was located on a side street, less than a mile north of the down town area. Hans parked in front of the store. A man in his thirties greeted them and said, "What can I do for you." Lena smiled at the man and studied the layout, keeping in mind that they might have to assume a defensive position. She saw three buildings; the home on the left, then a combination showroom and conference room and on the right was the work shop, where a number of gunsmiths manufactured various type of shotguns and rifles, including the infamous ray guns. She held out her hand, "I am FBI agent Lena McCabe, from Washington DC and I'm sure that you know Hans. He is helping us and he also serves as a tour guide."

"Of course I know him," said the man, "I attended his ski classes. I am David Franzel. My father is in the office waiting for you."

"Shall we go in?" said Lena.

"Of course." David led the way to the office. When Lena stepped in, she saw an office adorned with trophies of large and small game and awards for craftsmanship. Across the length of the far wall, was a build in gun cabinet, stocked with rifles, shotguns and yes, ray guns. The ladder is a new product that Franzel decided to manufacture, using his CNC equipment. In the middle of the room was a mahogany conference room table, a pot of steaming coffee and delicious assorted Viennese pastries next to it. Johann Franzel reached out and introduced himself. He was in his eighties, in excellent health, graying at the temples. It was obvious that he was a sportsman; he spent a large part of his life, traveling the world, hunting game.

"Please be seated," said Johann, smiling.

"Thank you, said Lena and continued, "I suppose you know why we are here."

"Absolutely. You are going to help me with the roboids. How could you possible allow them to get started, releasing dangerous bats and killing people?"

"We have no excuse for that. However, we are committed to put a stop to their impetus of trying to take over the world, before it's too late."

"I have heard what you have done the last time that you were here. I must thank you for that. And you are actively involved in other parts of the world to stop them. Am I right?"

"That is true. Do you know why the roboids are targeting your business, Mr. Franzel?"

"Because ray guns are a real danger to them?"

"Correct. One short beam will destroy their circuitry and render them useless."

"I never thought that our ray guns would ever be used for that purpose. However, this type of threat is not new to us."

"What do you mean?"

"Well, we almost got wiped out by the German forces during WW2. At that time, I was just a teenager. However, we convinced them that we could be useful to them by manufacturing implements of war. That's how we survived."

"Interesting. Let's get to the point." said Lena, "According to my sources, the roboids should arrive today. We do not know how they are going to make their entry into town and how they are going to be dressed. We must be on the lookout for any stranger entering the city. Hans will be of great

159

help pointing that out to us." Johann nodded, leaned back in his chair and said, "I think that you should go down town and wait for the roboids. I have a security system installed and I can activate it at any time, pointing ray guns and shotguns at intruders."

"If that's the case, we'll let you do your thing, Mr. Franzel. Thank you for your hospitality. We might be back later." Lena and Hans returned to the down town police station.

CHAPTER 31

Julius sat near the window at the police station, looking for unusual activity. He was surprised when he saw a Mitsubishi truck stop in front of the station. Lena and Hans left the truck and walked into the policer station, surprising Julius, "What's going on, Lena?"

"We have a change in plans. Did you see any strangers, yet?"

"Yes we did. One man checked into the hotel. He arrived with an expensive car and he has his skis mounted on a rack on top of the car."

"Really. Did you check him out?" said Lena.

"No we didn't." Lena glanced toward Hans, "Hans could you check out the person that just checked into the hotel."

"Absolutely. That's easy for me. After all I work there, part time."

"Put on you ski jacket and hide your ray gun under your jacket," said Lena, with authority, "If you have to use it, use it wisely. Whatever happens, we'll cover for you."

"I believe that you will." Hans Weider strapped on the ray gun holster and donned his ski jacket. Then he walked across the plaza and entered the hotel. In the middle of the lobby was a roaring fire; a couple of skiers sitting near it, with their boots supported from on the edge of the fireplace. He walked up to the clerk's desk. The desk clerk new him well, "Hi Hans. What can I do for you on this fine day?"

"Let me see your register. It's important." The clerk flipped the register around and Hans looked at it and said, "Boy its slow today. You have only one person that signed in."

"Exactly. He was a friendly, timid man. He signed in as Paul Dopfer."

"Did you see anything unusual about him? What did he do? Did he ask for something?" The clerk collected his thoughts, then said, "Yes, he ordered breakfast up to his room. I said that it is a little late for breakfast. But we *do* accommodate. I ordered it for him. I'm assuming that the maid took it up to him."

"I see. What did he do with his car?"

"Oh, yes. He asked whether he can park out front. He wanted to go skiing right after he ate." Hans nodded, "And what did you say?"

"I told him that normally he should park in the back, but we could make an exception with him. He thanked me. Then he left."

"Did he talk with an accent?"

"I think that he was from Germany. You know — the way they talk in Munich." Hans pressed his lips, nodded and was ready to leave. He was satisfied with the clerk's answers. He turned and walked right into a tall and extremely attractive female, "I'm so sorry. I hope that I didn't scare you, miss?" said Hans, holding her arm. She was close to him, looked up at him and smiled, "No you didn't." Hans smelled the expensive perfume and couldn't resist her, "Are you going to stay in the hotel?"

"Yes, I am." In the meantime, Hans let go of her arm. He noticed that it was solid, *she is definitely athletically inclined.*

162

"I am here to take skiing lessons. Do you know anyone her that might help me?" Hans was elated to no end; *would it be possible that I might become romantically involved with this beautiful woman?* Scanning her from top to bottom, he noted that she might be mixed and Hans knows how to classify beautiful women; 10% oriental, 15% black and 75% white. Would it be really his luck to meet a woman like this? With all these thoughts in his mind, he finally said, "Miss, I would love to teach you how to ski. I'm a professional ski instructor."

"Great. When can we meet?"

"How about at one this afternoon. I know of a perfect hill for beginners. By the way, my name is Hans Weider and what is your name?"

"You can call me Angela. My full name is Angela Barret." She reached in a small bag, pulled out a chocolate mint and ate it. Hans though, *bad breath, no way.*

"Good. I will see you right here at one p.m." Hans was elated. He walked across the plaza to the police station. Before he had a chance to close the door Lena said, "What did you find out?"

"A skier checked in to the hotel. His name is Paul Dopfer. He seems harmless. Then I bumped into a beautiful woman. She was looking for a ski instructor, …."

"Let me guess," said Lena, "You volunteered to teach her."

"Yes, I did. How did you know?"

"That wasn't too difficult to figure out. So what are you going to do?"

"I will meet her at one this afternoon."

163

"What about Mr. Franzel?" said Julius, frowning.

"He said that he can take care of himself. But, perhaps we should check up on him later today, anyway."

"And that would have to be, while you are skiing, right?" said Julius.

"Right." Hans walked to the window and looked out. With the exception of an old woman, carrying a bag of groceries, the plaza was empty. Hans looked at his watch. He had still time before his skiing appointment. He walked to the deli one block from the police station, ordered a sandwich and a soft drink. When he was finished he said to the clerk, "Put in on my tab. I'll pay you next week."

"Now, don't forget."

"No. I won't," said Hans, pouting. He has a tendency to forget his bill, but they always catch up to him. He looked at his watch again. Now it's time. He drove his VW truck to his apartment, donned his ski clothes and stuck his skis and poles in the back of the truck. Then he drove to the hotel to meet Angela. She was waiting, and he said, "Where is your equipment, Angela?"

"I do not have equipment."

"Well, let's go, then. We will have to rent yours. Would that be all right?"

"Yes. I am so glad that you are helping me, Hans."

"No problem." Hans drove to a beginner's hill. It is located south of Ferlach, at the foot hills of the Alps, reaching into Yugoslavia. At the chalet, he rented skis, poles and shoes for Angela. He helped her with the ski shoes; he tried to lift her leg, shook his head and said, "Boy we must have led in your

164

shoes." Angela quickly helped him putting on her ski boots, while she was smiling at him. Hans grabbed his equipment and in a few seconds he mounted his skis. He showed her how to walk on skis. Then he showed her how to walk uphill and how to control sliding downhill, using the snow plow. Angela smiled and apparently, she loved skiing. When she was near the bottom of the slope, she fell. Hans looked at her tight, yellowish ski pants, he couldn't help noticing her sexy posterior. He tried to help her up, but she was fast. She was back on her feet, at the blink of an eye, smiling at him, flashing her perfectly aligned, white teeth. She noticed that he watched her and she said, "Hans I like you. I noticed that you are watching me." Hans turned red and said, "I am sorry, but I couldn't help myself. You are just too beautiful."

"Thank you. You are not too bad yourself." Angela was a fast learner and it gave Hans pleasure, teaching her. Before long, they took the chairlift to the top of the hill and they skied down together, like professionals.

"If I wouldn't know better, I swear that you had previous training."

"What would you do, if I did?"

"I would look foolish, Angela."

"You should not. Just look at the rewards that you will be enjoying later." Poor Hans couldn't believe what he just heard. His mind went berserk and his adrenal started pumping, placing him in an extremely embarrassing position, leaving no additional room in his tight ski pants, while developing a huge erection, "Let's quit, and go to your room," said Hans, barely able to speak.

165

"Yes. Let us do that." Neither spoke another word. First they returned Angela's ski equipment to the chalet, while Hans threw his in his VW truck. Then Hans drove to the hotel and parked it in front. Angela took the lead, going to her room. As soon as she was inside, she closed the door and stripped, leaving only a pair bikini-type shorts on her body, her breast solidly protruding from her body. Then she ripped Hans' clothes from his body. By this time Hans developed a huge erection again and Angela couldn't help herself following her natural and God-given instincts. She engaged in oral sex until Hans screamed from pleasure. She looked at him and said, "Hans, sit on this chair, pointing to a stool in the middle of the room. Hand obliged, not wasting a minute. Angela straddled him, supporting her weight with her left foot, and slowly pushed down, allowing his penis to enter her body. It didn't take more than five minutes, before Hans exploded in her satisfying him to the point where he became dizzy and nearly fell of the chair. While he was still breathing heavily, he said, "That was wonderful, Angela."

"I'm glad that you liked it. What do you want to do now?"

"I have to go back to the police station. We are expecting four roboids. You wouldn't know anything about it. Would you?"

"No, I do not."

CHAPTER 32

It is later in the afternoon and Johann Franzel was getting tired. After all, he is eighty-three years old.

"I'm going to lay down for a while. You have to watch out for the roboids," he said, to his two grown children. While he slowly walked away, his daughter said, "We will, dad." She is pretty and she is two years older than her brother. Both of his children are concerned, especially after they saw what the roboids can do with their bats and their weapons, on the other end of town. David was on the way out and his sister said, "Where are you going?"

"I'm checking around the houses on our property, to make sure that we have all the guns in place."

"We should have our shotguns ready, just in case they bring their bats. But I have the feeling that they may not bring their bats to our home. They are after our ray guns. For that they might bring explosives."

"That's a scary thought." However, both felt that it won't hurt to be ready for any contingency. After they checked their home, the office and the gunsmith shop they were convinced that they did all that they could do. Three of their employees are positioned in various places, keeping an eye out for approaching strangers. Until now it was quiet. Could this be a false alarm? Who knows? The sibling returned to the office building and poured themselves a soft drink. Then they sat in a comfortable couch, waiting and hoping that the menace might go away. The office building is the original and the safest building on their property. David's great-grandfather had it built

167

from imported stone. David reached for a gun magazine. He turned to the last few pages and checked for the Franzel advertisement. It was a full-page ad, describing their weaponry, including their ray guns. One of the observers on the roof destroyed their mock tranquility. He initiated a short burst of a hand operated siren. The siblings jumped up and reached for their ray guns and they looked out. A VW truck pulled up in front. David recognized it, "It's only Hans Weider." He stepped outside the office and waved at Hans. At the same time, Johann Franzel came into the office from a back door. The siren woke him, "What's going on? Are the roboids here?"

"No dad. The roboids are not here. Hans Weider returned."

"That's all?"

"Yes, dad. That's all. You should be glad."

"I am, I am." Hans walked in and took off his ski jacket, exposing his ray gun. David saw it and said, "Are you cleared to carry this gun?"

"You are god dammed right. I'm cleared by the FBI."

"Really. Do they have jurisdiction over here?"

"Yes, they do. They are cleared internationally. Didn't you know that, David?" David looked embarrassed, "No. I did *not* know that." There was always a competitive spirit between them. He turned and looked out the window that faced to the front. Hans saw that and he said, "Is anyone watching the back?"

"Yes. We have two men on the roof, watching the back. Actually, we are checking all around the property."

"Good." Hans looked out again and he realized that it

168

was getting dark, "I think that you should turn on all the lights around the property." David's sister rushed and turned on all the lights that they had. Then Hans went on the roof of the workshop and looked down, around the property. He noticed that a couple of areas in the back, extending into a meadow, were not sufficiently illuminated. He returned to the office and said, "Do you have a couple of spot lights?"

"For what?" asked Johann.

"In the back of the property are a couple of dark spots," said Hans, troubled, "They should be illuminated better."

"Oh, well. I have men on top of the roof, watching. I could send one more man up there," said Johann, trying to humor Hans and keep him off his back.

"You are the boss, Mr. Franzel. I hope that you are right." Again, the member of the Franzel family gathered in the office area. Hans, on the other hand was nervous. Finally, he called Lena at the police station and expressed his concern.

"It seems that you have done everything possible to convince Mr. Franzel to get ready for the roboids. I don't believe that he fully understands the danger that he is in."

"I did, Ms. McCabe."

"How long are you going to call me Ms. McCabe?" Hans looked at the phone, *how do I react to this?* Then he said, "What do you want me to call you?"

"You can call me Lena, Hans."

"Alright, Lena. That definitely will be easier for me. By the way, have you seen any unusual activity in town, Ms. McCabe, I mean Lena?"

"No. Except a young, beautiful lady was here, looking
169

for you. I believe it was your skiing student."

"Really. She was looking for me?" said Hans, surprised.

"Yes, she was. After all, you are a good-looking stud."

"Is this a compliment, Lena?"

"Absolutely. Your student was panting." Hans laughed. "Women. Any other activity?"

"No, nothing else." Now Hans was beginning to doubt that the information that Lena received from Dallas was false and he said, "Are you sure that Pharos gave you the right information."

"I am positive, Hans." Lena was barely able to say the word 'Hans,' when she heard a terrific explosion over the iPad. Then she heard two other explosions in succession following the first, "What the hell is going on, Hans?"

"Stay on the line, Lena — I will find out." Hans drew his ray gun with his right hand, while he held his IPad in his left. He looked out to the front — nothing. Then he ran toward the back of the property. David followed. Hans saw in the distance three men running toward the downtown area of Ferlach. Hans shouted into the iPad, "Lena three men are running toward downtown. Obviously, they are the villains who set the explosives. Get ready for them. We are following them." Both Hans and David ran after the men. Hans assumed that they are roboids.

In town, Lena took charge, "Create a perimeter around the plaza with Hambrush's officers. Block the side streets. Do not let them escape. Julius, you stay with me. Hans and young Franzel are following the three roboids. When you see them, use your ray guns and disable them. Don't be afraid to shoot and don't waste any time."

"Right." Lena heard this response from the men. Lena and Julius stood outside of the police station, and said, "Julius, I do not believe that the roboids understand that we acted so quickly and already talked to the men at Franzel's estate. That type of logic is something that is not inherent in the roboid's programs. That might be our only advantage."

"I agree. I hope that they stay together."

"Right." The perimeter was established. Lena and Julius waited. It took no more than five minutes when three roboids came running and found themselves surrounded in the middle of the plaza. They stopped and looked around. The humans raised their guns, but hesitated, despite the instructions that Lena gave them. That gave the roboids a few extra moments to open fire. They unloaded their machineguns and sprayed thousands of bullets around the perimeter, hitting most of the men. Lena and Julius were the only two that they did not hit, because they hugged the ground and at the same time fired at the roboids. Julius hit the first one in the chest. His gun dropped and he stood motionless. Hans, behind the roboids, hit the second roboid in the head. Sparks flew from his structure and immobilized the second one. At the far end of the plaza a policeman was able to connect with the third roboid. He hit him in the belly, exposing hundreds of feet of copper wire and titanium body parts. Julius looked at Lena. She tried to get up, holding her right arm. Julius saw that, "What happened, Lena?"

"I sprained my wrist, going down. I am so sorry. Julius." Julius helped her up and took her to a medic in the police station. He applied an ice pack and told her to sit. Medics rushed to the officers along the perimeter. Out of the twenty-six officers, four died, three walked away on their own accords and an ambulance took the remaining officers to the hospital, to remove bullets from their bodies. Lena, Julius and Hans went into the police station to recuperate. They are lucky that they are still alive.

CHAPTER 33

Julius looked for a paper cup. Hambrush noticed and said, "What are you looking for, Julius."

"Water for Lena. I think that she needs it."

"Oh, I am so sorry. I should have thought about that." In a combined effort, the two men managed to get a cup of water for Lena. She smiled at their clumsiness, "Thank you all. That is so nice of you." Both new that she was humoring them. Julius still looked confused. Lena realized and said, "What's bothering you now. Julius?"

"Didn't you say that we should be expecting four roboids?"

"You are right. We saw only three. Where did they come from, anyway?"

"From Dallas, according to Pharos." Hans listened and said, "I have an idea. I can check at the hotel. Julius would you like to come with me?"

"Yes, that would be a good idea. After all, this *is* my problem. Lena, we should be back shortly." They walked to the hotel and Hans looked for the clerk. He wasn't at his desk. He rang the bell and the clerk came from a room behind the front desk. He said to the clerk, "How many guests signed in today?" The clerk checked in the register and said, Paul Dopfer and Angela Barret. Just those two."

"Give me the room numbers and a pass key."

"I don't know if I should do that, Hans."

"My friend, I have an FBI official from the United States here with me. You better give me a god dammed pass key." The clerk looked sheepishly at a stern looking Julius and slid the pass key across the desk, while he gave Hans the room numbers. Both walked to the first room. It was Dopfer's room. They entered carefully. The room was empty, and the bed was not slept in. On the table was his untouched breakfast. "That is strange," said Hans. Julius pressed his lips and said, "I don't think that this is strange, Hans."

"Why not?"

"This tells me that Dopfer was a roboid."

"Why would he order breakfast? Roboids don't eat — don't tell me, Julius, I know. He ordered breakfast to throw us of the track."

"Exactly. Look over there. He still has a charger plugged into the wall. He fully expected to return. But we killed him."

"That son of a bitch. He must have left the room earlier to meet the other two roboids."

"Good thinking. Hans. Now let's look in the other room."

"This must be Angela Barret's room." They went two rooms down the hall way. Hans carefully inserted the pass key into the lock and turned it. The door opened. The two men looked in. At the far end, Angela stood in a corner, a wire extending from her chest to a convenience receptacle in the wall.

"Motherf...... Sorry Julius. The bitch is a roboid and I made love to her earlier today. And it was wonderful." Hans thought about past events, "But she ate candy, Julius."

"Yes, they are getting smarter and they have new ways to deceive humans."

174

"How is that?"

"I believe that she has a small chamber in her abdominal area. Food will accumulate there and when she is alone she can use a vacuum pump to clean out the chamber."

"I'll be god dammed." Hans kept staring at Angela. She was still standing motionless. He said, "What's wrong with her, Julius?"

"I believe that she powered down."

"I see." Hans went closer to her. Her eye lids were half closed and flickering slightly. Makeup was smeared and some of it was running down her cheek. Hans felt sorry for her. He gently touched her arm. She opened her eyes and shook slightly, straightened out and said, "Hans, what are *you* doing here. You caught me by surprise."

"You are a roboid, Angela."

"Of course, I am. What did you think?"

"But, I made love to you. I thought that you were human."

"No I am not. Let me show you." First Angela unplugged, then unbuttoned and unlatched her breast plate, with her ID information inside:

SN: F 97,436

Date of Mfg: 2/02/2016

Place: Dallas, TX

Roboid Name: Tania

Human Name: Angela Barret

Programmed for: Companionship

Languages: English, German, Italian

Hans showed the plate to Julius and he said, "I have

seen a plate like this before. She is harmless, programmed for companionship."

"I'll be god dammed. Now I have seen everything."

"No, not quite. Put the plate back on her chest and tell her to button up." Tania watched and said, "What should I do now?" Hans looked at Julius and he shrugged his shoulders, "What do you want to do with her, Hans?"

"I know what I would like to do with her."

"Really. Then keep her. She could be your property and she would cost you nothing, barring a small bottle of machine oil to keep her parts functioning smoothly."

"Are you out of your mind, Julius? I can't be tied down now." Hans kept staring at Tania, then said, "Why would they send *her* to Ferlach, when their objective was to eliminate ray gun manufacture?"

"You haven't figured that out have you?"

"I guess not."

"They sent her along to keep guys like you under control."

"But that hasn't worked for them, did it?"

"Fortunately for all of us, no, it didn't." Now, Hans was pleased with himself and he said, "Oh, well I guess I could keep her." He turned to Tania, "Grab your stuff and let's go." Tania fetched her grip and place her charger in her grip, "I am ready." Hans stopped by the desk clerk and threw the pass key on the desk, "Both rooms are empty. You can rent them out again, but you should clean them first." Then the threesome returned to the police station.

"Where were you so long?" said Lena.

176

"We checked out the hotel. Julius explained to Lena in great detail what they encountered.

"This was quite a story," said Lena, "Now let's walk to the Franzel estate."

"Good idea. I still have my truck there, we can drive back," said Hans.

"Right." They walked slowly. Lena had her right arm in a sling. Tania kept looking at Lena and she noticed, "Why do you keep looking at me, Tania?"

"I think that you are very brave and beautiful."

"Thank you. You are also beautiful."

"I was designed that way. I am programmed to please men."

"So am I, Tania. But please do not tell Julius that."

"I believe that I understand. The human female is designed to reproduce the human species."

"Yes that is correct. And if we do not attract men, our species might die out."

"That is very unfortunate. I am attracting men but I cannot reproduce the roboid species."

"That is true. Your species is being manufactured and occasionally upgraded."

"The human species is also occasionally upgraded — especially the female of your species."

"Yes. It is. Well here we are. You do not have to tell anyone that you are a roboid. Especially since roboids blew up the Franzel complex."

"I will not say anything." The group walked into the office. Johann Franzel and the siblings sat on their dusty

177

furniture. No one spoke and their spirits were broken. They just finished checking the damage of their property. The explosives destructed much of their property. The office was the only building that wasn't demolished, though the pressure of the explosives broke most of the windows. The factory part had a wall missing and their home needed to be rebuilt. In short, the place was a mess.

Lena started, "What are you going to do, Mr. Franzel?" Johann looked at his two children and said, "First we will rent a recreational vehicle, where we will sleep. Then we will rebuild the shop. That is the most important part. At the same time, we will order new windows for the office. Then we will find a builder to build us a new home."

"That is a good plan," said Julius, "How long will all that take?"

"We should be done by year end. I hope by tomorrow we will start selling off a few of our weapons and in one month we will start producing new ones, including ray guns."

"Well Mr. Franzel, we wish you the best of luck," said Lena, compassionately.

"And what are you going to do next?"

"We are returning to DC for new orders."

"Well, we wish you good luck and thank you for your help," said Johann. They went to Hans' VW truck and he drove them back to the police station.

A day later, Hambrush ordered a front end loader to scoop up the roboids. The operator loaded them on his truck. Later, a local artist, with the help of a paleontologist, reassembled them and built a museum around them on the

178

plaza, remembering the roboid attack in the year of 2016. They also included the tremendous losses that the Franzel family incurred on the same day.

CHAPTER 34

Christopher Barony tapped the enter key to break off the communication with Antony Marino. They sent messages to each other for nearly two hours, at the rate of over two-hundred words per minute. From the drift of their statements it was that evident Marino now has assumed a leadership position; that is continuing the impetus of destroying the human race. They agreed that Marino would expand his factory underground, building roboids at a record pace, while Barony would continue to supply the Ebola carrying bats, since he is already on the African continent. They agreed that planting bats in smaller communities would be the best approach to eliminate the human race. They have learned from their test sites what they must do improve their success rate, concerning both, the frugal placement of bats and the preservation of roboids. Both are also working on a masterplan to systematically destroy the human race. However, they also realize that they must preserve a selected group of humans for their needs. For example, Ramon Bergen must be allowed to build roboids for domestic use. Eventually, these roboids would be used as a cover to hide Marino's roboids. The Bergen roboids would ultimately be reprogrammed for population control and only a few would be safe from program alterations. Marino and Barony predict that in two to three years they would have control over the industrial centers on earth and at the same time reduce the population in third world countries. Roboids rationalized that natural resources on earth must be preserved, since materials to build roboids come from earth. During Barony's communication with

Marino, he stood by the elevated desk, a structure that a human architect designed. It serves the human need to provide comfort. Of course, as far as Barony is concerned, it does not matter if he stands or if he sits. Sitting is a position that humans prefer, because they get tired standing too long in one place. That is why they have on the one hand, an elevated desk, where they stand and on the other hand a regular desk, where they sit. Roboids also understand, when both, standing and sitting become cumbersome for humans, they lay down and rest and in most cases it turns into sleep, a condition which they equate to their recharging state. After they disconnected, Barony was not too happy with the drift of the conversation, because he always thought that he was the leader, however, he decided to play along for now.

Barony has made it his primary objective to study earth's humans in his spare time. He scanned the internet, studying the development of the human species. Any information that he can gather, above what he has now, will help him in the future to improve his understanding of humans. He found that humans are a weak species. He realized that they are not from this earth and that earth might be a prison planet for them, since humans are by nature a violent species. Humans cannot stay in the sun too long, the way lizards can. They cannot jump too high, run too fast and stay under water for too long. However, they have a superior brain, capable of programming roboids. This is the area that Barony is concentrating on. After having studied the development of humans, organizing known and conjectured facts, Barony came to the conclusion that humans evolved somewhere else. Then

181

certain aliens transported them to earth, about 80,000 years ago. And since that time, these and other related aliens have been watching humans developing into the species that he sees today. They have improved in many ways, but physically they are still the weakest species on earth. Barony loaded another 100 GB of information into his computer, to continue his studies, when a man stepped into his office and said, "Am I correct in assuming, that you are the police chief of Tunis?" Barony looked at the man, dressed casually and immediately realized that he is a roboid, "Yes I am the police chief. What can I do for you?"

"I am looking for a new home."

"Why do you come to me?"

"I have heard that you might be looking for additional policemen."

"What is your name?"

"My name is Baruk."

"Where do you come from, Baruk?"

"I escaped from the roboid farm in New City. It is operated by Ramon Bergen."

"I see. I have heard of him. How did you get to this continent?"

"I changed my appearance. Then I withdrew money from three ATM's. Then I acquired a passport on the black market and bought a plane ticket. Then I boarded a plane in Dallas and flew to Tunis. Then I ran to this police station."

"Very ingenious. What are you programmed for, Baruk?"

"I am programmed for both, domestic work and police work."

182

"Have you had any experience in you programmed field?" In other words, have you been able to activate your self-learning chip?"

"No sir. I escaped before Bergen sent me to my first assignment."

"I see. I tell you what. Report to the sergeant behind me. He will task you."

"Thank you, sir." Baruk turned, while looking of the sergeant and reported to him. The sergeant gave Baruk a temporary job which was exactly what Baruk was hoping for.

"Clean out all the rooms in the police station, including the large auditorium," said the sergeant. With this job, Baruk was able to study the design of the building, which included the layout of the station.

Right after Dactor cleaned up the city of Qued Ellil of bats and roboids he stored most of his implements of destruction at the police station. He did this with the help of the sergeant in charge. In his subprogram, listing car dealerships, he found a second hand dealer that sold trucks. He picked a Mercedes truck, with an oversized suspension system, paid for it and drove it to the Qued Ellil police station. Then he loaded dynamite, plastic explosives, clamps and related accessories on the truck. He also included an ample supply of wireless electronic detonating devices. As an afterthought, he added four, full five-gallon cans of gasoline. Next he secured a solid steel cover over the cargo bed of the truck and locked it. He returned to the police station and fetched his grip with personal belongings and went to the locker room to change. He cleaned up and then he donned a tailored, gray leather suit, accented

183

with a white shirt and a striped tie, when the sergeant walked in. "Wow, you look terrific, Dactor."

"Thank you. Starting now, I am a roboid leader." The sergeant placed Dactor's spare clothes in the grip and zipped it, "I must say you *do* travel light. Sometimes I am actually jealous of you." Dactor grabbed the grip and threw it in the passenger seat. The sergeant waited for him by the truck, "Good luck Dactor. I hope everything will proceed according to your plans."

"I have to admit, sergeant, right now I do not have a plan. But I have a good idea how to proceed. I will be depending a lot on Baruk." Dactor signaled his facial muscles to produce a smile that tried to radiate sincerity. Dactor stepped into the truck and left. Since it was only twelve miles to Tunis it took him only twenty minutes to get there. He parked the truck in front of the Tunis police station and went in. A tall policeman stopped him and said, "What is your business here?" Dactor looked at the man and he realized that he was just doing his duty and said, "Let me pass. I am here to see Chief Christopher Barony."

"The chief is busy right now. Can you come back later?" Dactor became impatient. He gently pushed the policeman aside and started to walk in. The tall officer grabbed Dactor by his arms and pulled him hard. Dactor, being programmed as a leader, with unsurmountable strength, grabbed the officer around his waist and threw him against the wall, bending a number of titanium parts and severing a main conductor, "You are going to the repair shop, buster." Barony heard the clatter and he came running, "What the hell is going on here?"

"Are you Barony?"

"Yes, I am. And I am also the policed chief of Tunis."

"Well, I am the new leader. You can stay on as police chief, as long as you follow my orders."

"Who said?"

"Marino, said. He sent me here to investigate. He wants to know why so many roboids terminated, costing Marino a fortune. And why you botched up all the test sites."

"I am going to call Marino. He is in charge, not you."

"If you call, he will deny everything that I said. Besides, he is in Dallas and I am here. Now let us go to our office." Barony headed toward his office. Abruptly, Dactor heard the klick of a sear on a pistol, he turned and saw that Barony pulled a gun, ready to shoot him. However, Dactor was expecting that Barony was going to shoot. He had his ray gun ready and when Barony turned, Dactor fired his ray gun, splitting his chest plate in half and frying the cables behind it. Barony remained standing and Dactor called for a few officers to clean up the debris that he left behind and said, "These two are going to the repair shop." Then he went to the office that Barony occupied and sat behind the desk. He called in all the officers on duty and told them that he is now in charge. He looked on the bulletin board and studied the existing schedule for all three shifts, and said, "I see that you have about one hundred policemen controlling the city of Tunis. Do you think that this is enough?" One of the officers stepped in and said, "We could use more men on third shift. We are spread too thin. We have a difficult time controlling the human drunks."

"I see. What other problems do you have on third?"

185

"Right now, most crimes occur on third. Humans know that we do not have enough policemen on third."

"I see. I thought that you will be engaged in population control. That will considerable reduce the number of humans. Am I right?"

"Yes, Dactor, you are right. But Barony did not initiate a movement toward that goal."

"He just cleaned up the test sites. Did he not?" said Dactor.

"Yes, he did. Perhaps I am assuming too much."

"Yes, perhaps you are. I suggest that you return to your jobs, until I will have a chance to evaluate the whole situation. For now, I will reassign two roboids to third." The officers left, but Baruk remained. Dactor looked at him and said, "Stay here, I will load a new program for you." Baruk opened the chest plate and Dactor inserted a memory stick, while Baruk patiently waited to be upgraded per Dactor's specifications.

CHAPTER 35

Pharos stood in front of a large table, reviewing the original drawings that Marino used to build this vast underground factory, as well as the new drawings for the expansion of the complex. Krista was sitting by the desk, processing new orders for the roboid repair shop, as well as shipping memos for the finished product. She moved slowly and nearly powered down. He looked at Krista, noticing her lackadaisical approach to her work and said, "How many roboids ae you going to ship today?"

"Today I am going to ship four of them, Pharos."

"What will be their destination?"

"A small town near the outskirts of New York City. Marino said that this will be his new test site."

"I see. Hold this shipment until tomorrow. I will give you new orders, later."

"Yes, Pharos."

"Get going, Krista. Plug in and recharge."

"Yes, Pharos."

Pharos looked busy, however he knew that looking at the plans for a major expansion was busy work. He had other plans. These plans are the reason why Lena McCabe assigned him to Marino's roboid repair shop five weeks ago. During this time, Pharos was busy repairing and manufacturing roboids, while Krista was shipping them to various destinations. During the same time, Pharos was also busy determining the key points where he would place explosive charges, designed for maximum effect, to bring the manufacturing floors down. He determined that the primary locations would have to be the many support columns which support the ceilings. He marked these places on his drawings. He avoided the main house and the areas where the horses grazed. To make the expansion plan more plausible, he ordered an underground mine drilling machine with spare cutters, including a loader, a truck and ceiling support beams from a West Virginia mining company. Every day, early in the morning, armed with one of his ray guns, he stopped by his Ford truck, and took a couple sticks of dynamite to the manufacturing area. Inconspicuously, he placed them near the base of various support columns, and he fused with wireless ignitors. He cleverly hid them inside the void areas between the steel columns and the outer casings. At the same time, one of his roboids operated the mine drilling equipment, starting to clear a large space, where the new manufacturing area was going to be, while another roboid placed temporary support columns.

Pharos walked through the complex one more time, reassuring himself that he placed all the explosives correctly. While he was in a secluded area of the property, he pulled out his communications device and called Lena McCabe. After a time, longer than normal she said, "This is Lena, Pharos. Why do you call me at such ungodly hour? Never mind, you can't sleep. Right." Pharos instinctively knew that Lena was joking. He decided to answer with a joke, "I was sleeping very well. In fact, I was dreaming." Now Lena was at a loss of words. Finally, she realized that Pharos is beginning to understand human jokes, "Alright, Pharos, what is so important?"

"I am ready to blow up the complex, including the

owners."

"Great. What about you?"

"What *about* me?"

"What are you going to do?"

"You will find out after I demolish the place. I will leave their home standing. And the horses will roam freely in the pasture."

"Good luck, Pharos."

"Thank you." Pharos disconnected. With the ray gun and the remote control hidden from view, he went to Marino's home and said to Antony and Lexia, "I am ready to show you the progress of the factory addition." Antony was delighted. He believed that Pharos made tremendous progress and soon he would be able to double his production of roboids. All three walked through the empty animal husbandry building toward its end. Then Pharos operated the elevator that took them down to the factory. Antony and Lexia followed Pharos to the end of the factory. They stepped behind a temporary wall. Antony saw the mine drilling equipment. It was obvious that he was disappointed and he said with a raised voice, "That is all that you did during all this time?"

"Not quite. But right now, I am going to a lot more." Pharos pulled his ray gun and shot Marino in the chest. He never had a chance to react. Then he aimed at Lexia and shot her also in the chest. Pharos still saw the two structures shaking, sparks flying and smoke escaping. Then he pulled his remote control from his pocket and tapped the enter key. About twenty-five explosions occurred simultaneously, which included the elevator shaft. The whole area shook, approaching earthquake

189

intensity, with a magnitude between 8 and 9 on the Richter scale. This massive explosion created a sinkhole of about 2 acres in diameter. The farmhouse remained standing, however one wall collapsed. Two horses dropped into the sinkhole, but the rest of them managed to escape, jumping over a collapsed section of the wooden fence, neighing and screaming. They escaped to the far reaching pasture. Lexia, Antony and Pharos expired. Pharos blew the Marino Roboid Repair Shop into oblivion. And that included himself.

CHAPTER 36

The Tunis Police Department is located on a two-floor structure and it is one of the oldest building in Tunis. On the first floor is the main desk, designed to keep a comfortable distance between the sergeant in charge and people entering to resolve their businesses. Behind the front desk is a long hall way, with detective offices on both sides of the isle. Right now, most of them are empty. On the far end are two toilets and a kitchen. Both of these are now rarely used, except if a human needs to use one of these commodities. Upstairs are sleeping quarters, as well as storage rooms. Baruk has studied the layout of all these room, ever since Dactor changed Baruk's program. He was busy and went to work. He pushed all bunks to one end of the room, keeping one bunk for humans, strictly as a courtesy to them. In the basement are numerous jail cells. Right now, only one human was locked up, awaiting a trial. Baruk ignored the human, though he was screaming and complaining, that he was unjustly handcuffed and jailed. The day before, he stopped in a pharmacy and took three bottles of alcohol. He hid them in a shopping bag and promptly walked out with it. A clerk, behind the counter pushed an emergency button and a few seconds later a roboid from the Tunis police department showed up, running at seventy miles per hour. The clerk told him that the customer was stealing. The roboid apprehended the thief and jailed him. Actually, the roboid did not know why he acted the way he did. A subprogram directed him to react to the concept of stealing. Driven by a subprogram, he reacted accordingly.

While Baruk studied the layout of both floors of the police department, he paid particular attention to the auditorium. He noticed that the seats are arranged, slanting upwards. Being a roboid, Baruk ran this information through his processor and it responded noting that humans could see better if the row behind the row in front of the row is slightly elevated. Baruk tried to accept this perception, though the message was a weak facsimile of this simple concept, that humans invented. Next, he continued, studying the layout of both floors. When he completed his work, he met with Dactor and reported his findings to him. Since Baruk did not know the reason for his study, he did not list the load carrying beams, both hidden and visible. Dactor decided to tell him that he is working on a plan to abolish all roboids working at the police station. Baruk's self-learning chip kicked in and he said, "What will happen to me, Dactor?"

"I will try to save you, but most likely you will be a part of the demolition process. However, if you can come up with a method to abolish all roboids, while you could save yourself, let me know." Baruk's cameras flashed for a millisecond and then said, "I will come up with a solution to save myself, but only if I can stay on and work for you, providing that you are going to stay and run the police department."

"That is my intention, Baruk." Being satisfied with the answer, he left for the auditorium to further study the layout and the load bearing uprights. When he was finished he returned to Dactor and showed him his findings. Dactor stood by the front window, looking at the plaza. When he saw Baruk, he said, "Go to my truck and fetch dynamite and detonators. Attach them to

192

all the support beams in the police station on all floors. Make sure that no one sees you. Sometimes you roboids are very careless. And that is what gets you into trouble."

"But you are also a roboid, Dactor."

"Yes, I am. But I am a leader, with much greater strength and responsibility."

"Right. I can do what you ask for, Dactor. If you do not mind, I would like to replace the detonators with a wireless variety. Then I could ignite them, while I am outside the police station."

"A good subprogram variation. Do it and at the same time you can save your presence."

"Thank you. I appreciate that. When are you going to assemble the police force and have your meeting?"

"Tomorrow at nine. You have to gather them all up from all three shifts. Make sure that you place all the charges in time. Work all night, until you are done."

"I will do that." Baruk left. He walked to Dactor's Mercedes truck. He removed the steel cover and looked in and said, "Holy smoke." He rationalized, *I should have no trouble blowing up the police station with all the dynamite and gasoline that is available to me.* He donned a loose-fitting work jacket and hid dynamite sticks under it. Then he went to work, placing dynamite sticks and wireless ignitors on all the support beams on all floors. Late at night, stretching into early morning, he took one five-gallon gasoline can to the basement, two to the first floor and one to the attic. He placed two ignitors on each can. When his job was done, he reported to Dactor and he told him that he completed the job. Dactor asked about the gasoline

193

cans and Baruk told him that used all four of them.

"Good job," said Dactor. "Gather all my policemen from all shifts and tell them that I will be changing department policy. And that the meeting will be at nine in the auditorium. Can you do that Baruk?"

"I sure as hell can, Dactor." Dactor smiled.

The same day, at nine in the morning, Dactor stood on top of the podium. He scanned the police force and counted — one hundred and four. He was convinced that he had all of them in the auditorium. Then he said, "Excuse me for one moment, I forgot to bring my charts, describing my new policy." He swiftly ran to the front of the station and signaled Baruk to tap the remote control.

Baruk tapped the button on the remote control.

At first it was quiet for two seconds. This short span of time seemed like an eternity. Then all hell broke loose. The roof separated from the building and flew upward about twenty-five feet, then it came crashing down. At the same time all four walls of the station blew outward, accompanied by furniture, pieces parts of roboids, toilet fixtures and weaponry that was stored in gun cabinets in the office of the police station. Flames shut up thirty feet into the sky, caused by the gasoline. One roboid came staggering from the rubble. Dactor saw him and said, "How the hell did he survive." He pulled his ray gun, aimed at the roboid's chest and fired. The chest plate separated, cables sparked and the roboid fell forward right on his face.

"Perfect job," said Baruk, smiling at Dactor.

The Tunis police station was gone. With exception of Dactor and Baruk, no one survived. Dactor pulled his

194

communications device, tapped Lana McCabe and said, "The Tunis police station is gone. What do you want me to do now?

"Get a cleanup crew and start rebuilding."

"All right. I will do that."

CHAPTER 37

Lena finally had a day of leisure in her FBI office. It was a pleasant, clear day and she felt good about herself. Her gun was hanging from the coat rack and she had no intentions of carrying it today. She was on the road for over four weeks chasing roboids and bats, from sand covered roads in dire need of repair, to meadows covered with snow, sparkling in the sunlight. Perhaps this would finally be a good day to go out with Julius for breakfast. She called him, hoping that he would be in his office. After three rings, he replied, "Hi Lena, don't tell me that we have another emergency."

"No, no, nothing like that, Julius. I'm taking you out for breakfast, across the street."

"Stop by my office on your way down." Julius closed his laptop and, straightened out his desk and donned his jacket. He grabbed his overcoat, then looked out the window. He saw two men leisurely crossing the street, down below. They wore their suits, no overcoats. Julius pressed his lips and decided to do the same. He returned his overcoat to the coat rack and waited for Lena. Five minutes later, Lena peeked in his office and said, "Are you ready?"

"As ready as I will ever be." She wore her business suite and she also decided to go without an overcoat. Julius noticed, "No overcoat?"

"No. It's a beautiful spring day."

"Right. Not too long ago we went skiing."

"Wow, how time flies."

"Yes, that's what my mother used to say." They crossed

the street, after they waited for the green light as well as a hand waiving them on. When they entered the restaurant, a waiter said, "Good morning Ms. McCabe. How are we today? I haven't seen you in quite a while."

"We are fine," said Lena, winking at Julius. The waiter placed them by their favorite table and he said, "The usual?"

"Yes Jim, the usual." Finally, the agents relaxed and sipped their coffee. Lena looked at her IPad.

"Are you expecting a call?" Lena smiled, shaking her head, "It's a stupid habit. It's the job, Julius. Even when I'm on vacation, I wonder who will call next, sending me to a remote part of the world."

"I understand. We are married to the FBI; on call twenty-four hours of the day."

"Right. So, now that we have established that, how are we going to close this case?"

"Do you really think that this case will ever be closed?"

"Martin might deactivate it temporarily. But if there were another report of folks dying and bones crumbling, you can be assured that we would be back on the road."

"Exactly," said Julius and continued, "Tell me, what will happen to the property in Dallas? By the way, who does it belong to?"

"I believe that now it belongs to the state. We should let Martin deal with that part."

"That would be a good idea." The waitress stopped by with a fresh, steaming coffeepot in her hand, "More coffee anyone?" Lena shrugged her shoulder, "Yes, why not. Who knows when we will have the luxury having another breakfast

197

on a Friday morning?" The waitress poured coffee and left. Both agents sat silently, but it was obvious that their minds were racing. Are they done with the roboids? What has Martin Hall in store for their next project? And how soon will that happen?

"I know what you are thinking, Lena."

"What am I thinking?"

"You are wondering if we will be able to build roboids in the future and have total control over them."

"Yes, something like that. I believe that we must review this with Ramon Bergen. Make sure that his roboids are totally controlled by humans."

"Definitely. How do you feel about Dactor, Lena?"

"Dactor will be alright. When I talked to him the last time, reviewing the stability of his program, he said that he will be busy hiring humans for the new Tunis, police force."

"How about the police station?"

"Martin told me yesterday evening that the Tunisian president will be aiding Dactor, building a new police station."

"Good." Lena waived for the waitress and paid her. Then Julius and Lena rose and slowly returned to the J. Edgar Hoover, FBI building. When they walked past Martin Hall's office, he said, "Where the hell were you. I have scheduled a meeting at eleven a.m. I need for both of you to attend." Julius looked a Lena and said, "Shit. Didn't I tell you?"

"Yes, you did."

www.ingramcontent.com/pod-product-compliance
Lightning Source LLC
Chambersburg PA
CBHW070229210526
45168CB00019B/286